D0459436

SPACE PLACES

PHOTOGRAPHS AND TEXT BY
ROGER RESSMEYER

FOREWORD BY BUZZ ALDRIN

THIS BOOK
WAS MADE POSSIBLE THROUGH THE
GENEROSITY OF

Eastman Kodak Company

Nikon

IBM

COLLINS PUBLISHERS

For Jain,
who shares in my exploration of life on this planet,
and for those who have lost their lives
in the exploration of space.

First published 1990 by Collins Publishers, Inc., San Francisco, California

All photography provided by Starlight Photo Agency. Any inquiries regarding the photography may be directed to the agency.

ISBN 0-00-215732-2

Library of Congress Cataloging-in-Publication Data
Ressmeyer, Roger, 1954 -
Space Places / by Roger Ressmeyer; foreword by Buzz Aldrin.
1. Outer space—Exploration. 2. Astronomy. 3. Photography—Journalistic. I. Title.
TL790.R47 1990 520-dc20 90-1483

Printed and bound in Japan
First printing April 1990 10 9 8 7 6 5 4 3 2 1

Previous page:
Dawn's early light over the McDonald Observatory on Mt. Locke, Texas, U.S.A.

C O N T E N T S

F O R E W O R D

I am convinced that somewhere in space there is a Sutter's Mill that will lead to the stellar equivalent of a California gold rush. I don't know where this mine is, nor exactly what it holds, but I am certain it is out there. Its "gold" may lead us to new technologies; it may take the form of access to natural resources in the asteroid belt; it may even take the form of contact with extraterrestrials. But wherever and whatever "it" is, the promise and possibilities are as infinite as space itself. Most important, the prospects for humanity's survival may well depend on finding it.

So how, and when? In America, our decision-making bodies are besieged with a wide array of problems, so there is a natural and understandable tendency to rely on conventional wisdom and old "sure" ways of doing things to narrow the options. The problem is that such approaches promise slow, if any, progress. NASA is today being criticized for a "business as usual" attitude, when what is required is bold innovation.

I am also concerned as to who will explore space. The primary current barrier to America's commitment to space exploration is cost; the cost of getting payloads into low-earth orbit alone is exorbitant, currently running in the high thousands of dollars per pound.

But that is a transient problem that will be overcome. Every means of transportation has gone through a cycle of high initial investment succeeded by steadily reduced costs until it is available at low price to everyone; space will be no different. The jet airplane, which began as an expensive military tool, has become the transport of choice for millions of people. Space transportation may take a bit longer; but if we wish, it can become common transport also. Once the means of space transportation is developed, it will be a constant.

The effects of cheap space transportation on civilization, however, will be much more far-reaching than earlier advances. Cheap access to space will move us out into the solar system frontier, leading to unlimited raw materials and energy. Combined with the access to information created by the computer revolution, this will give us the three elements necessary for unlimited wealth: energy, raw materials, and knowledge.

So how do we do it, how do we begin this crucial exploration? To begin with, man must be in space. Machines are not adequate to the task. All they can find are the things they have been programmed to look for. But space has too many unknowns, and presents too vast a frontier. Robot space probes are simply not capable of fully assessing and tapping its resources.

As a first step toward manned exploration, today's simulation technology can allow us to explore different options in a fashion not available in the past. Computers are now powerful enough to develop complete mission scenarios. We should therefore use them to "field-test" entire missions, weighing competing technological and strategic options against one another before deciding which to launch.

Look back. As we developed the world, we had to build seaports, then airports. Now is the time to recognize that we will have to build orbiting spaceports. Space makes movement easy. And wandering space stations will give us a freedom and a low-energy, continuing-mission capability that was never available to other means of transportation. Short-range space transports will shuttle back and forth between these ports and the planets. I'm mystified as to why there has not been more of an attempt to utilize such moving bases in space, and wonder why we don't use this concept as the gateway to this next great venture.

As a former astronaut, I also find disturbing the apparent lack of a long-term national commitment to space exploration. If Mike Collins, Neil Armstrong, and I had suffered tragedy on our moon-landing mission, that lack might be understandable. As the Challenger explosion demonstrated, it's difficult to keep the dream alive in the wake of disaster. But we've overcome that setback, and Apollo 11 was indeed a resounding success. As John Kennedy envisioned, the Apollo Program demonstrated to the world just what people are capable of accomplishing when they set their minds to it.

It therefore seems time to stop the short-term programs. Now we must go out there for the long haul. It's easy to ask, "Why explore space now, when we have so many problems here on Earth?" But that question will never go away. Yes, we can always wait for a better time and a better way, but then we will do just that: always wait!

Clearly, the essential ingredient for getting deep space exploration off the ground is the understanding and support of the public; without it, politicians won't appropriate the necessary funds. With it, space is ours.

Apollo 11 photo by Buzz Aldrin – NASA

I'd like to think that Roger's images can help. Certainly they will create a deeper understanding and appreciation for the grandeur of the concepts and the truly epic proportions of the opportunities humanity has created with the opening of the space frontier. To quote a much used truism, "a picture is worth a thousand words"; to that I would add, "and a great picture is an essay."

Beyond that, Roger has placed the space challenge into context with other great human achievements. In doing so, he has allowed us to experience the vast flow of human development and growth, to see our opportunity in space as the next great step in an open-ended voyage into infinity—an inevitable step, not just of human beings traveling to new worlds, but of the human spirit blossoming toward capabilities that we, even now, can only dimly glimpse. Enjoy this journey of his, and let us hope it represents just the beginning of our future in space.

—Buzz Aldrin

Like buried treasures, the outposts of the universe have beckoned to the adventurous from immemorial times....

Princes and potentates, political or industrial, equally with men of science, have felt the lure of the uncharted seas of space ...

Enterprise

And through their provision of instrumental means the sphere of exp

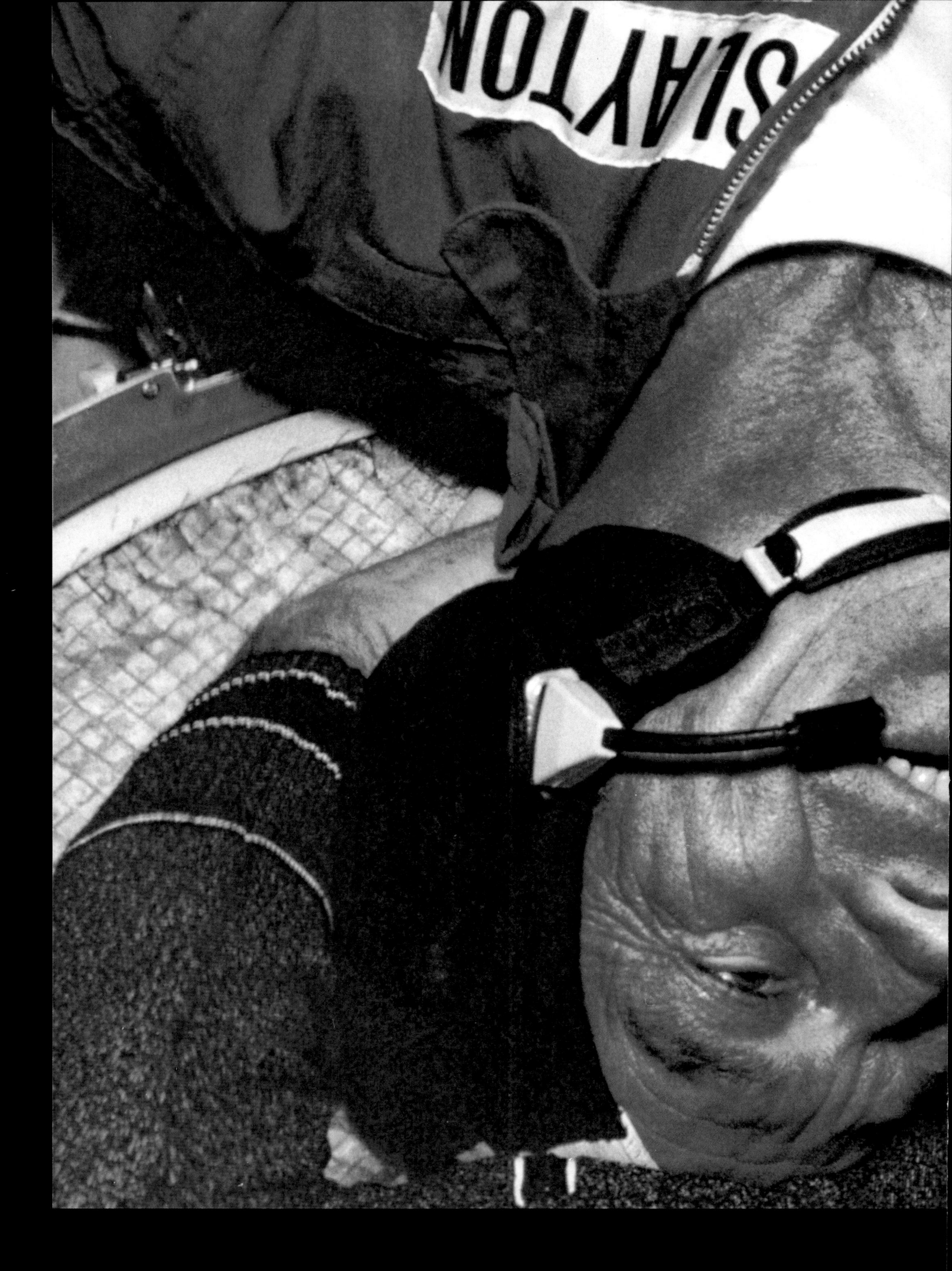

Each expedition into remoter space has made new discoveries ...

Apollo-Soyuz photo by Thomas Stafford – NASA

And brought back permanent additions to our

The latest explorers have worked beyond the boundaries of the Milky Way in the realm of 'island universes'....

Apollo 17 photo by Eugene Cernan – NASA

While much progress has been made, the greatest possibilities still lie in the future. —George Ellery Hale, 1928

I N T R O D U C T I O N

Ever since childhood, space exploration has enthralled me. As a boy, I liked its mysterious implication of the infinite and the eternal. As an adult, my career as a photojournalist has enabled me to visit the world's leading centers of research—space places, I call them.

I have wanted to be an astronaut ever since my eighth birthday, when I discovered that there was a place above the blue sky called space, and that an American named John Glenn was about to visit there. That dream, I suspect, was one shared by an entire generation.

The 1960s were an ideal time to be a schoolboy interested in space. One astounding discovery followed another in the span of just a few short years. While I was flying model rockets, building telescopes, and reading science fiction, astronauts and space probes were mesmerizing the world. As I looked at the moon through my first telescope, I joined scientists in trying to imagine what people would find there. The moon's surface was totally unknown. Could it swallow our landing craft? Might there be life buried in its mountains? Yet by the end of the decade the moon had become an old friend. The world had lived through the most thrilling period of exploration in history: Live, from the moon, Neil Armstrong and Buzz Aldrin!

I have yet to join the astronauts, but I've already lived in their world: flying with them in weightlessness-simulating jets, diving with them into neutral-buoyancy pools, and delighting in the stunning glory of their launches. Seeking to understand humankind's ageless yearning for the stars, I've also traveled to observatories atop the darkest mountains in the world—a magnificent experience. And along the way I've come to realize that while the roots of our odyssey into space are older than recorded history, we're only now beginning to realize its rewards. For the same technology that takes us to the stars is exactly what's required to ensure the quality of life on earth.

Many of the secrets to preserving our tiny, delicate planet lie near at hand, if only we can learn to recognize them. Around the world, space people are working on critical problems that precisely mirror the environmental challenges facing the planet as a whole. It's no coincidence: A lonely spacecraft hurtling through the solar system is simply a microcosm of life on earth, with all the same needs. Techniques for the safe production of energy, for recycling and waste management, for finding natural resources, and even for ensuring cooperation between nations are all necessary elements of any long-term voyage through space—and all are directly applicable to real-life situations around us.

There are still those in America who say we shouldn't waste money on space when we have problems at home. Yes, space research is costly. It won't solve the quandaries of drugs, poverty, crime, and despair. But it can make significant contributions. Space exploration inspires nations of people to achieve the exceptional, to bring cutting-edge technologies into everyday use, and thus to improve the world as a whole.

For example, the majestic photos of Mother Earth taken by Apollo astronauts fueled the environmental movement by helping people realize just how small and fragile our beautiful home really is. We looked at manned spacecraft with their limited supplies of fuel, food, and water, and saw an analogy to earth, in that long-term survival depends on renewable resources. We studied the globe from orbit and saw no boundaries between nations. We found direct proof that we are destroying our atmosphere and oceans. That awareness has helped lead to action.

Space research can also contribute to an energy-rich future for our planet. Unless we solve the world's energy problems before we run low on fossil fuels, technological society may be forever lost. To do that, we need to understand solar fusion so we can safely duplicate it on earth. We need to consider the possibility of recovering the helium-3 deposited on the moon by the solar wind, since it is the best and cleanest fuel for fusion. We need to investigate the potential in earth-orbiting solar power stations. And we need to improve energy efficiency so our renewable resources will go farther.

Communications satellites have already revolutionized international relations. But space exploration has had a profound effect in a more subtle way. Over centuries, we've gradually come to see that humanity doesn't occupy the center of the universe. As our importance in the grand scheme of things appears to shrink, the inevitable realization takes hold that we must take care of ourselves, of our world, of our resources, of our environment, and especially of our neighbors on this fragile planet.

Just as space flight offers a biological model of life on earth, so too does it provide a highly visible stage for the resolution of political conflict. Partners in any space venture are obliged to make serious commitments to one another, and every day the astronauts themselves invent new ways of overcoming obstacles to international cooperation. Place ten individuals from around the world in a single spacecraft bound for Mars, and that group will quickly evolve into a peaceful and cooperative unit. All nations will inevitably grow close through the experience of watching the travelers learn how to understand and communicate with each other.

It is time for the world to lay down its weapons, and learn to share precious resources rather than fight over them. The final frontier lies within man himself—in his maturity, and in the health of the relationship between each and every member of the human race. Space is a place where we learn to confront our limitations in a very direct way, and where we might first conquer our instincts for hatred and destruction.

These feelings are shared by many of the people I've met while wandering the globe in search of space places. Space explorers, and I include earthbound astronomers among them, are bright, committed, and serious. They are dreamers, visionaries. They are people who can help move the world forward into a healthier tomorrow.

Over the past twenty years, as I made the photographs in this book, many of these wonderful people helped me by explaining and sharing the special worlds they inhabit. Many times these same people worked with me to make the unusual and mysterious appear on film. Often this meant meeting at three o'clock in the morning, or keeping a team of technicians on hand until midnight. At other times, their research had to be interrupted. I am grateful to all of them.

Roger Ressmeyer in NASA's KC-135 during free fall.

None of my pictures are fakes; no darkroom tricks were added later. When the camera's shutter closed, after a fraction of a second or after ten hours, the image was rendered on film exactly as it appeared through the lens. Sometimes during long exposures I "painted" the scenes with light, my brush strokes discreet bursts of illumination in the corners of these spacescapes. Many noteworthy places I visited are not included in this book. But I have tried to touch on all the key elements of our journey toward the stars.

To gaze into space is to embark upon a spiritual quest, an experience of awe and wonder, a longing for the farthest horizons. Today it is time for the human race to chart a bold course into the 'Second Space Age', establishing a permanent presence in the sky, an interplanetary economic system, and a new era of expanded international cooperation. This book reflects my voyage of discovery into this awakening second space age. Welcome to what I have found.

—Roger Ressmeyer

Photo by David Malin – ©1981 Anglo-Australian Telescope Board

Exploration is really the essence of the human spirit. —*Frank Borman, 1969*

ANCIENT ASTRONOMY

The roots of space exploration go back to the very dawn of civilization, a time when night skies everywhere were brilliantly lit with an awesome panoply of stars. Thousands of years before Magellan circled the globe and proved firsthand that it was round, people were already observing the effects of planetary rotation and orbital revolution. The swirling and weaving motions of the sun, moon, and planets aroused wonder and confusion because basic concepts of orbital motion were not understood. Instead, the complex movements were attributed to temperamental gods.

The earliest sky watchers could at least seek peace of mind through prediction of those movements, and the most obvious place to begin was with the seasons. Using the North Star as a point of reference, observers in the Northern Hemisphere could chart the progress of the sun's path from north to south and back again over the course of a year.

With the coming of winter, sunrises and sunsets occurred farther south along the horizon. As days grew shorter and temperatures fell, ancient observers saw that the sun slowed and then stopped its southward motion, and began moving north again. The ability to predict the onset of spring gave a sense of order to an otherwise perplexing world.

Other celestial events, though—eclipses, comets, meteor showers, novae, and northern lights—were cause for alarm because of their unpredictable timing. And the wandering motions of the planets Mercury, Venus, Mars, Jupiter, and Saturn were especially puzzling.

Astronomy began as an effort to fathom these motions and events. Over tens of centuries, instruments were created in Asia, Europe, Africa, and America that could plot positions of stars and planets with increasing precision. With these measurements, Aristotle, Ptolemy, Copernicus, Kepler, Galileo, and Newton were able to construct theories about earth's place in the observable universe, and to conclude that the sun, not the earth, was at its center.

Today, positional astronomy continues under a new name—astrometry—and we understand that the sun isn't the center of the entire universe either. But the centuries of painstaking research and fervent speculation that laid the foundations of this increasingly precise science can still best be appreciated at space places like those on the following pages.

The Great Pyramids
Giza, Egypt

Built around 2,600 B.C., these imposing structures were aligned with Thuban, the North Star of that era. But Thuban, in the constellation of Draco, the dragon, is no longer precisely overhead at the earth's North Pole. Over centuries, the earth wobbles on its axis like a slowly spinning top. It was the degree of the pyramids' misalignment from Thuban that helped modern archaeologists date the period of their construction.

Stonehenge
Salisbury Plain, Great Britain

Construction at the prehistoric temple of Stonehenge began over 5,000 years ago and continued for 20 centuries. The largest stones were erected around 2,000 B.C., some weighing over 40,000 kilograms (45 tons) and standing nearly seven meters (22 feet) high.

The axis of the inner stones was aligned precisely with the sunrise on the longest day of the year, the summer solstice. On that day, observers sighting from the center of the circular monument could watch the sun rise directly over a distant pillar called the Heel Stone. Over the centuries, the sight line to the Heel Stone was periodically reoriented to compensate for the earth's wobble. Other stones and sight lines marked positions of the sunset and sunrise throughout all of the seasons, though the summer solstice sunrise seems to have held unique significance.

Even today, the magic of that special day persists. Thousands of visitors try to invade Stonehenge for the dawn on that morning, requiring road closures and police protection to guard the fragile stones.

Some archaeo-astronomers suggest that Stonehenge was also used to predict eclipses of the sun and moon. Although the observations of the Neolithic and Bronze Age people were not sophisticated by present standards, there is little doubt that Stonehenge was, at least in part, one of the first true astronomical observatories.

Ulug Bek Observatory
Samarkand, U.S.S.R. ●

By the 15th century A.D., astronomers were capable of making detailed measurements of celestial positions. Mongol prince Ulug Bek's facility (above) included a 40-meter marble arc with precise markings. A sighting platform was slid along the arc to record the positions of stars and planets when they crossed an imaginary north-south line in the sky.

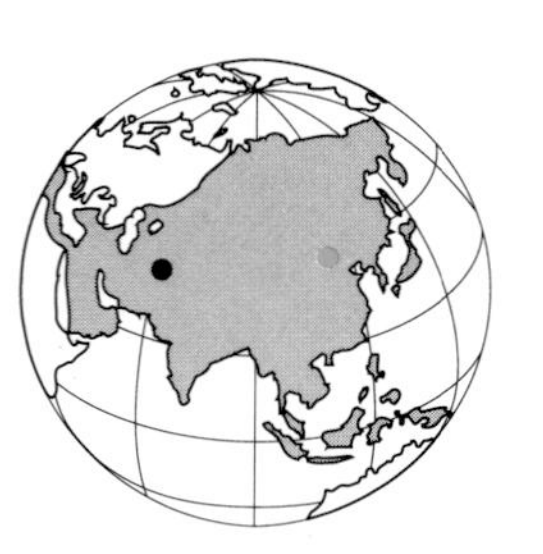

Ancient Beijing Observatory
Beijing, China ●

Increasingly sophisticated instruments enabled Chinese observers to track the stars and planets in all directions, not just along the north-south meridian. The armillary sphere at right, called the New Armilla, was completed in 1744 A.D. It was used to measure true solar time as well as "right ascension" and "declination"—the celestial equivalents of longitude and latitude.

The New Armilla features a sighting tube, seen here pointed at the moon, mounted within a sphere consisting of several rotating rings. Other astronomical instruments are positioned nearby. The Ancient Observatory was built in 1296, not far from today's Tiananmen Square.

Pulkovo Observatory
Leningrad, U.S.S.R.

Observatory director Victor Abalakin celebrates the 150th anniversary of Pulkovo, the world's oldest major observatory still in use. Destroyed during the second World War, it was later rebuilt. Here the science of positional astronomy has been continually refined, using telescopes to improve the precision of celestial measurements.

The largest optical telescope at Pulkovo, seen at left, has a lens 66 centimeters (26 inches) in diameter. It is used to measure the orbits of multiple-star systems, where two or more suns revolve around each other. Pulkovo telescopes are also used to study solar flares.

Telescopes ... cannot be so formed as to take away the confusion of rays which arises from the tremor of the Atmosphere. The only remedy is a most serene and quiet air, such as may be found on top of the highest mountains above the grosser clouds. —Sir Isaac Newton

Ever since 1609, when Galileo first pointed a tiny hand-held telescope at the sky and discovered the moons of Jupiter, humanity has steadily pushed back the barriers of space and time through the development of larger and better instruments.

The past century has witnessed a virtual building spree of these giant light buckets, fueled in the 1920s by a series of profound discoveries. American astronomer Harlow Shapley theorized that the sun lies at one edge of the Milky Way, a galaxy made up of perhaps four hundred billion suns, and that the Milky Way is but one of millions of galaxies, which he called "Island Universes." No longer was the sun, or even our own galaxy, regarded as the center of creation.

Edwin Hubble then showed that the galaxies are literally racing apart from one another. His evidence was the "red shift," whereby black lines in a galaxy's rainbow-like spectrum shift slightly toward the red end in proportion to the speed of recession. In the same way, the sound of an approaching train's whistle drops to a lower pitch as it passes and then recedes from a listener.

The study of galaxies, which are dim objects under the best conditions, kindled a craving for bigger and better telescopes. These pages explore the milestones of this development. Today a second wave of telescope construction is just beginning. In the coming century, advanced technologies will lead to the construction of far larger instruments as well as to new orbiting observatories.

The purpose of a modern telescope is to gather, concentrate, and focus light from distant objects onto film or electronic detectors. Light-gathering ability is directly related to the diameter, or aperture, of the primary lens or mirror through which a telescope funnels light: the bigger the aperture, the more light collected. A ten-meter (400-inch) telescope, for instance, has an aperture four times larger in area than that of a five-meter (200-inch), and will gather four times as much light.

But even the collecting power of large apertures can be diminished by local sky conditions. A small telescope in orbit can see farther into space than a larger earth-based one compromised by nearby city lights and dense, polluted air. Only where the air is thin, steady, dry, clean, and dark is the "seeing" considered good.

Modern astronomers have responded to light pollution and poor visibility by placing their telescopes on high mountains, far from civilization. These dark, clear sanctuaries are among the most beautiful places on earth, and the starlit skies above them an awe-inspiring sight, even to veterans. In this section we tour the great telescopes, beginning with the oldest and finishing with those still not completed.

Meudon Observatory
Paris, France

The problem of light and air pollution is well understood at the Meudon Observatory near Paris. The advent of street lights and smokestacks has dramatically impeded viewing conditions since the facility was founded in 1876. Once the home of Europe's largest telescope, the observatory today studies the sun during daylight hours when light pollution is not a factor.

Lowell Observatory
Flagstaff, Arizona, U.S.A. •

The great American telescopes of the late 1800s were mostly "refractors," which used primary lenses from the renowned Cambridge optical firm of Alvan Clark & Sons. The 61-centimeter (24-inch) telescope at near right, delivered to Lowell Observatory in 1896, was Clark's last.

Lick Observatory
Mt. Hamilton, California, U.S.A. •

In 1888, a 91-centimeter (36-inch) Clark refracting telescope was installed in the world's first permanent mountaintop observatory (far right). For years Lick Observatory offered supreme seeing conditions ... until the city of San Jose materialized in the valley directly below it.

Refractors use primary lenses made of two pieces of perfectly transparent glass polished on all four surfaces. By contrast, a reflecting telescope, like the three-meter (120-inch), 1959-era instrument seen in the largest dome in this picture, uses a curved, dishlike mirror to gather light. Only the front surface of the mirror is ground and polished; then it is coated with reflective metal. Because no light passes through, imperfections within the glass are irrelevant. The mirror can also be supported from behind to prevent sagging and distortion. For these reasons, the largest telescopes in the world are now reflectors.

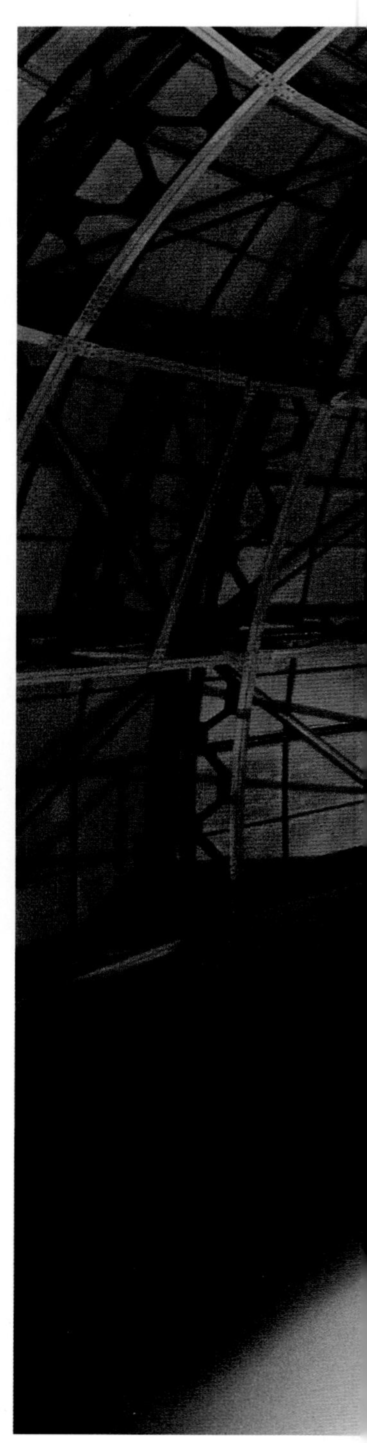

Mt. Wilson Observatory
Mt. Wilson, California, U.S.A.

From 1908 until 1949, Mt. Wilson was the home of the world's largest reflecting telescopes, the first measuring 1.5 meters (60 inches) and the second 2.5 meters (100 inches) in aperture.

The "Hundred-Inch" Hooker Reflector (right) was completed in 1917 after six years of tedious work grinding and polishing the 4,100-kilogram (9,000-pound) plate glass mirror to within two-millionths of an inch of a perfect parabolic curve.

Mt. Wilson's studies of distant galaxies were the rage of the 1920s and 1930s. Edwin Hubble used the Hooker to measure the velocities of and distances to faraway galaxies, thereby proving Harlow Shapley's contention that our Milky Way is only one of millions of galaxies, each consisting of billions of stars. His observation that nearly all galaxies were receding at high speeds from the Milky Way led later astronomers to theorize that a "Big Bang" created the expanding universe.

Located on a mile-high peak a few miles north of Pasadena, the observatory enjoyed exceptional sky conditions during its early decades. Later, lights in the San Fernando Valley brightened the sky, spoiling the visibility of dim galaxies.

Spiral Galaxy, NGC 2997

(Following pages) One of the most important astronomical developments of the 20th century was the discovery that "spiral nebulae" were actually galaxies of stars rather than clouds of glowing gas, and therefore far larger and more remote than previously realized. This galaxy in the southern sky is similar in structure to our own Milky Way.

Photo by David Malin – ©1980 Anglo-Australian Telescope Board

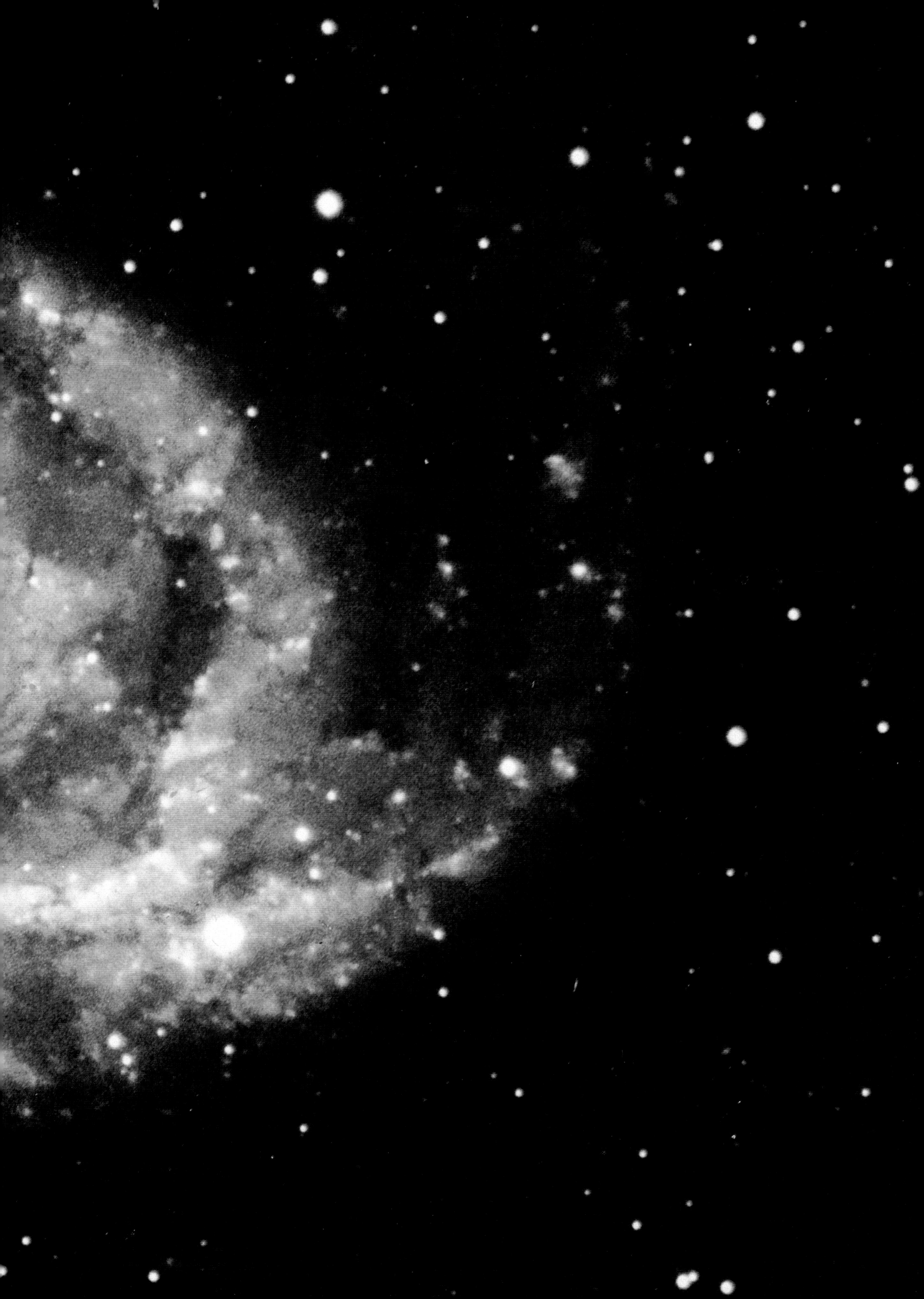

Palomar Observatory
Palomar Mountain, California, U.S.A.

In 1949 the five-meter (200-inch) Hale Telescope on Palomar Mountain surpassed Mt. Wilson's 100-inch as the world's largest. George Ellery Hale was the visionary behind all three great telescopes at the Mt. Wilson and Palomar Observatories in Southern California, including the one that bears his name.

The Hale has been one of the most productive telescopes of all time, and it will certainly be used well into the 21st century. Urban development in nearby San Diego and Los Angeles has brightened the sky at Palomar Mountain, but astronomers have enlisted the help of local governments to counter the problem. Street lights in neighboring communities now use low-pressure sodium lamps, and in some areas the lights are shielded as well. The limited wavelengths emitted by those sodium lamps are easily countered by light pollution filters at the telescopes.

For wide-angle viewing, an observer sometimes sits within a capsule suspended at the front end of the telescope's tube. Incoming starlight passes around the capsule to the main mirror, which reflects it back to the prime focus inside the capsule itself. The 200-inch made its last photographic image of the heavens from this capsule in 1989. In the future, all observations will be made with advanced electronic light detectors rather than film.

Every few years the Hale's huge mirror is cleaned and stripped of its reflective coating. The pyrex disk is then placed in a vacuum chamber where aluminum is vaporized at high temperature. The aluminum vapor condenses on the glass in a layer just a few millionths of an inch thick, creating a brilliantly reflective surface.

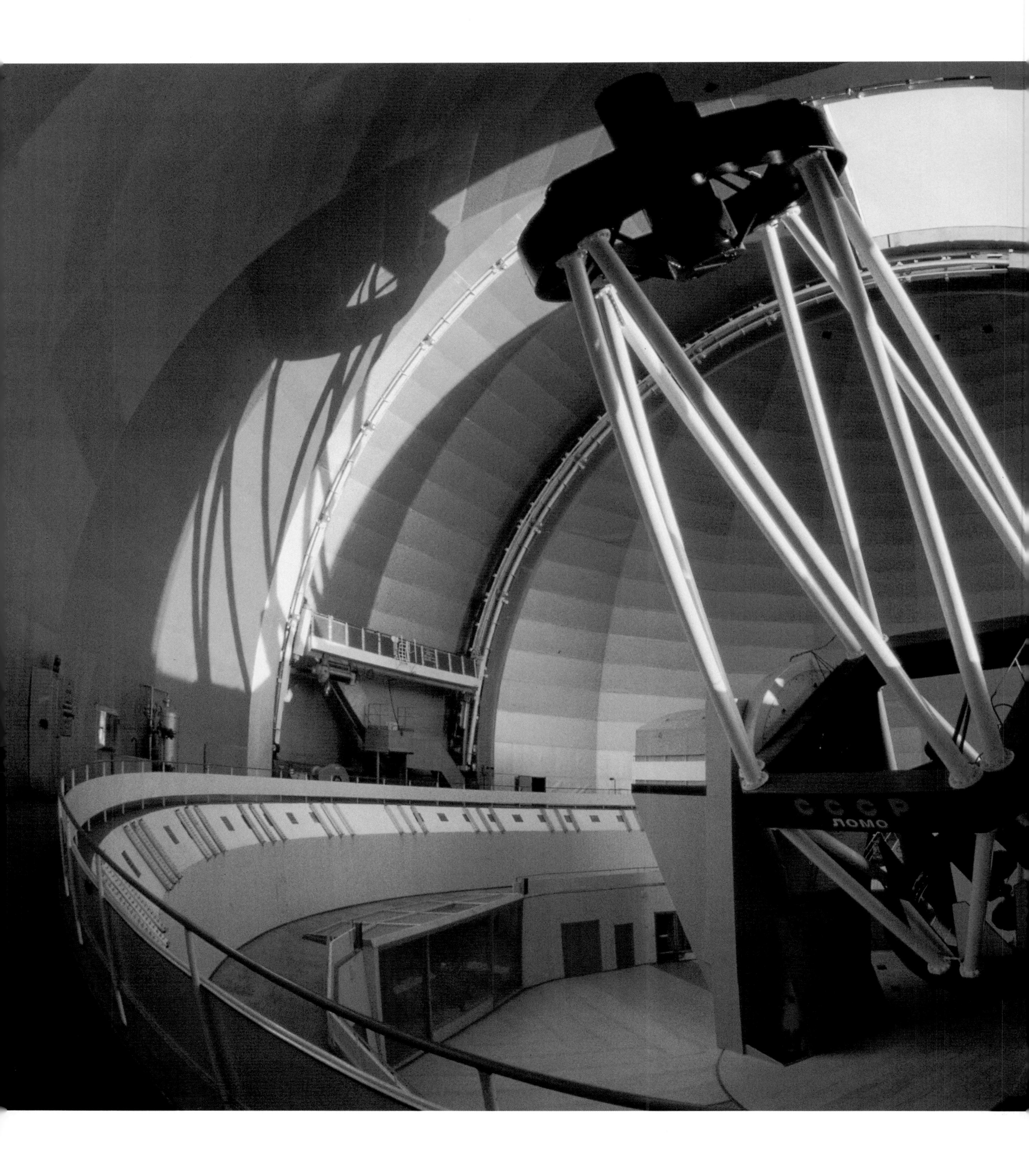
СССР

Special Astrophysical Observatory
Mount Pashukovo, U.S.S.R.

Completed in 1976, the six-meter (236-inch) Bolshoi Alt-azimuth Telescope, or BAT, is currently the world's largest optical telescope. Located near Zelenchukskaya in the northern Caucasus Mountains, this reflector uses a simplified telescope mounting that resembles a gun turret. The pioneering design saves weight and mechanical complexity by using a computer control system to correct for the earth's rotation.

Much of the mass (and expense) of older telescopes is found in their equatorial mountings, which are tilted to allow one axis to stay parallel to the earth's axis. A clock drive keeps pace with the earth's rotation by turning the telescope about that tilted "polar" axis once every 24 hours. The earth's spin is thus magically negated, and the view made stationary.

By contrast, the BAT's up-down, left-right (alt-azimuth) motions do not mechanically follow the curved paths of the stars as the earth turns. Instead the telescope's computers accelerate and decelerate the telescope at continuously changing rates about both axes, while simultaneously spinning the instrument mounted at the focus.

As a result of such sophisticated computer control, the Soviet six-meter mounting is much lighter and less expensive than the equatorial mounting on Palomar Mountain's smaller five-meter telescope.

In the 1990s, the BAT will be succeeded by larger alt-azimuth telescopes. In Hawaii the 10-meter Keck Telescope is already nearing completion. In Western Europe planning is underway for a "Very Large Telescope"—the VLT—that will collect the light from four separate eight-meter telescopes into a single image with the equivalent brightness of one produced by a single 16-meter mirror. Japanese engineers are designing a 10-meter instrument. And Soviet scientists have announced plans for a 25-meter monster to be deployed on Sanglock Mountain in Central Asia.

The BAT has been criticized for its location and optical quality. But while the site is not as ideal as Mauna Kea and Spain's La Palma, it is high and dark, and far better than Palomar Mountain. The main mirror has had problems, but it is still doing useful astronomy.

The greatest problem faced by Soviet astronomers at the BAT is the inability to procure sophisticated electronic instrumentation. Lack of "hard" international currency in their budget has precluded the purchase of advanced light detectors built in Western Europe, America, and Japan.

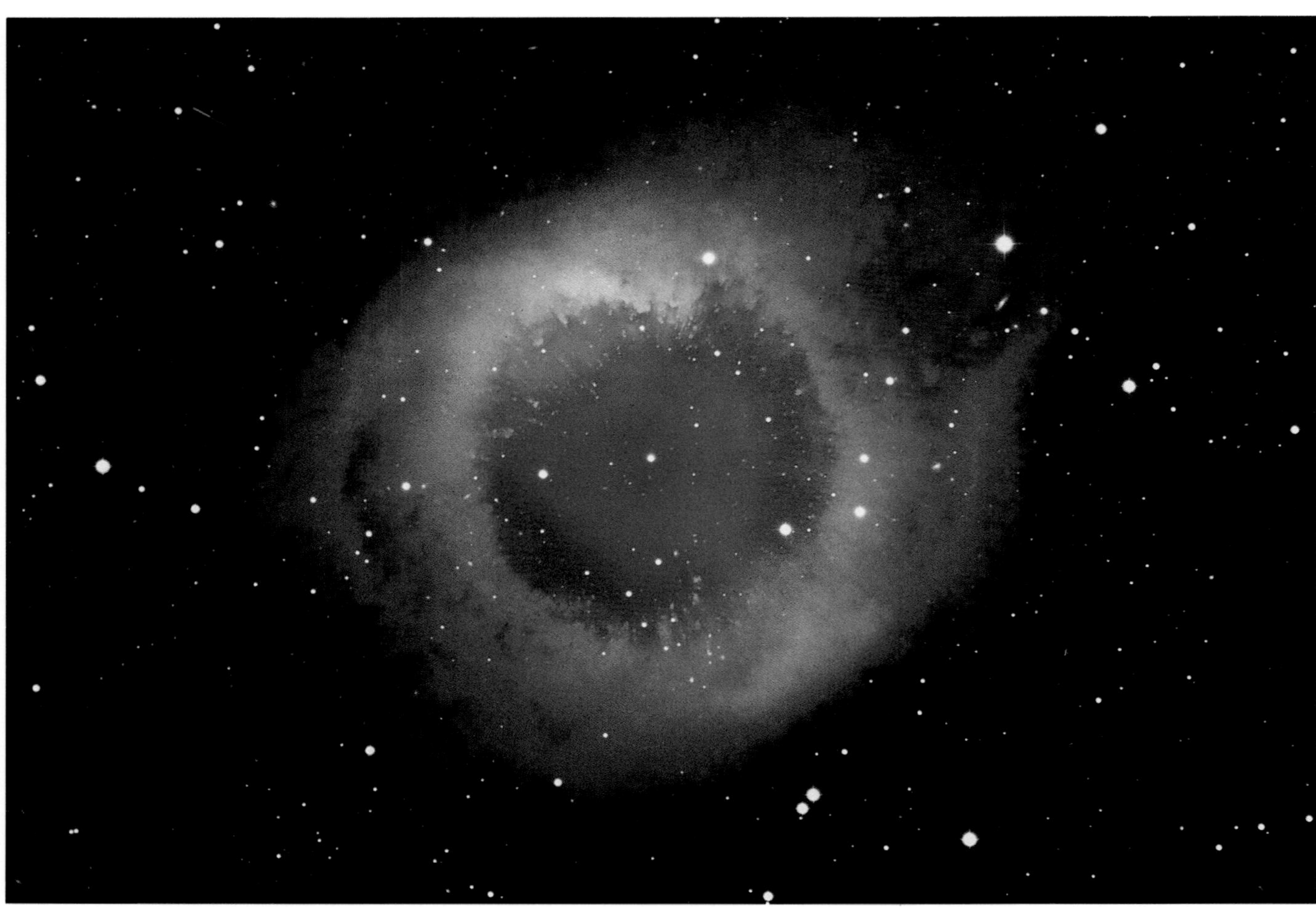

Photo by David Malin – ©1979 Anglo-Australian Telescope Board

Anglo-Australian Observatory Siding Spring, Australia

(Previous pages, and right) Since its completion in 1975, the four-meter Anglo-Australian Telescope has been used by David Malin to produce some of the most beautiful astrophotographs of all time.

Sitting in the prime-focus cage, Malin guides the huge telescope while exposing each subject on three separate monochromatic plates. By combining these green, blue, and red pictures, his finished prints portray the true colors of deep space with absolute accuracy.

Malin uses an additional photographic technique called "unsharp masking" to enhance the detail in under- and over-exposed regions on his film.

Helix Nebula, NGC 7293

(Above) Called a planetary nebula because of its spherical shape, this cloud of glowing gas was ejected by a star that became unstable in its old age. The light from this object, traveling at a speed of almost 300,000 kilometers (186,000 miles) per second, took almost 400 years to reach earth.

Horsehead Nebula in Orion

(Following pages) The shape of a horse's head in this David Malin photograph is the result of light-absorbing dust that overlaps a cloud of glowing gas.

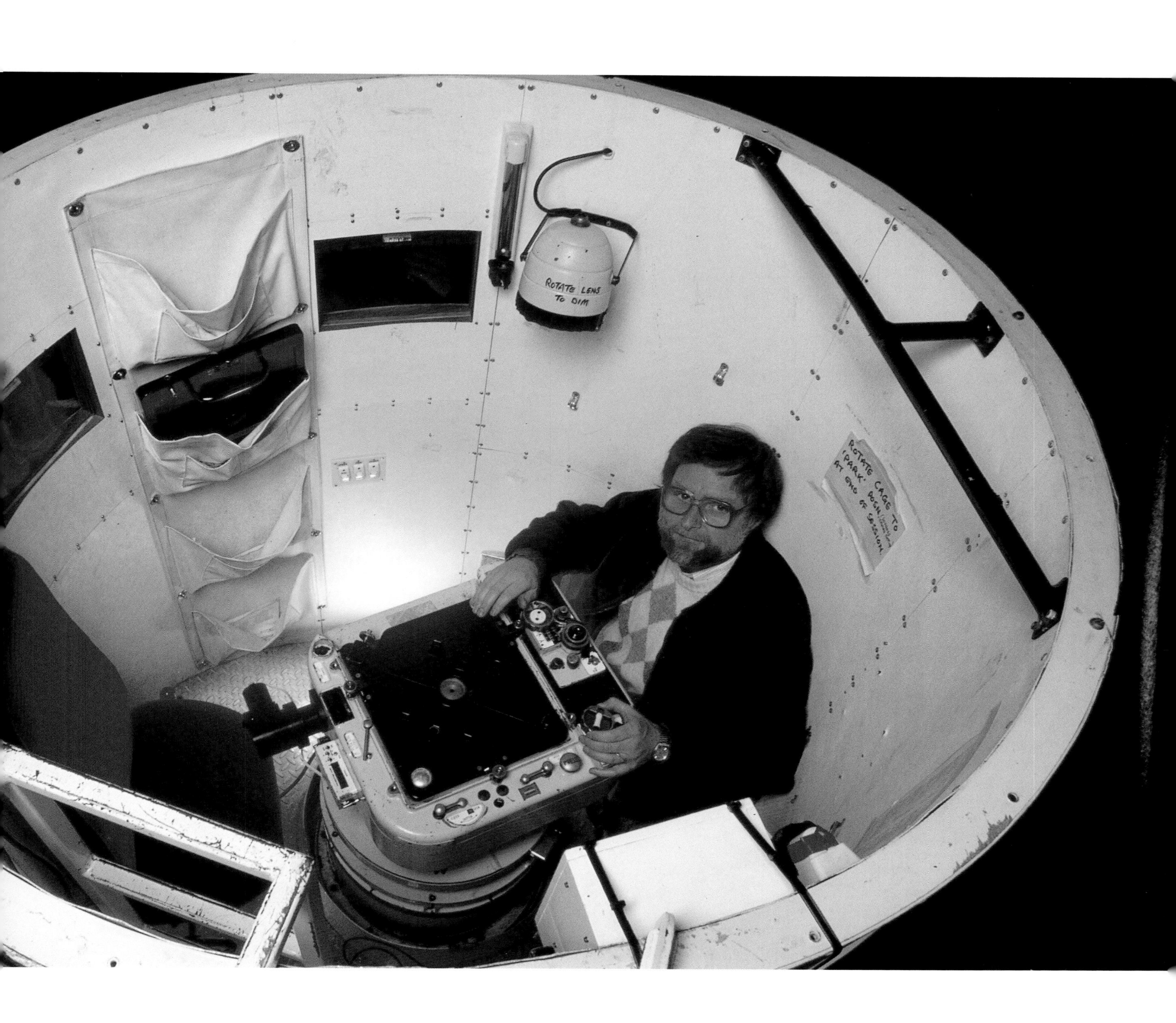
ROTATE LENS
TO DIM
ROTATE CAGE TO
'PARK' POSN
AT END OF SESSION

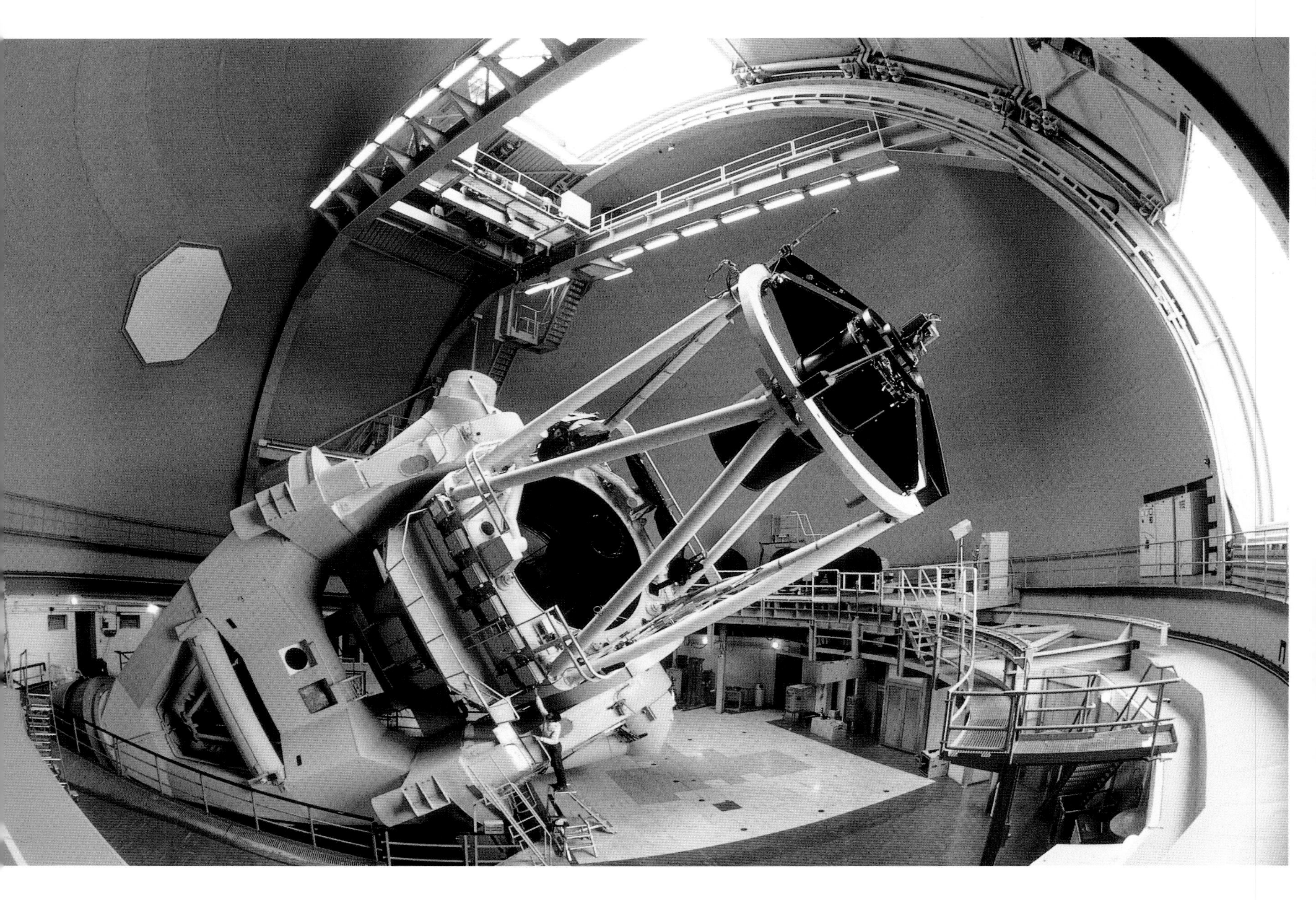

European Southern Observatory (ESO) La Silla, Chile

(Previous pages, and above) Modern observatories like ESO, in Chile's coastal mountains west of the Andes, are located where they face little competition from city lights. They are always situated atop high mountains, where the air is thin and clear.

At ESO, the European Community has assembled a collection of large telescopes to scan the southern skies, picking out objects that can't be seen from the Northern Hemisphere. The largest instrument in opèration to date is the 3.6-meter, equatorially-mounted telescope shown in this photograph. Recently a second 3.6-meter telescope was commissioned at ESO. Called the New Technology Telescope (NTT), its purpose is to test concepts for the VLT, the 16-meter instrument ESO plans to install in the 1990s.

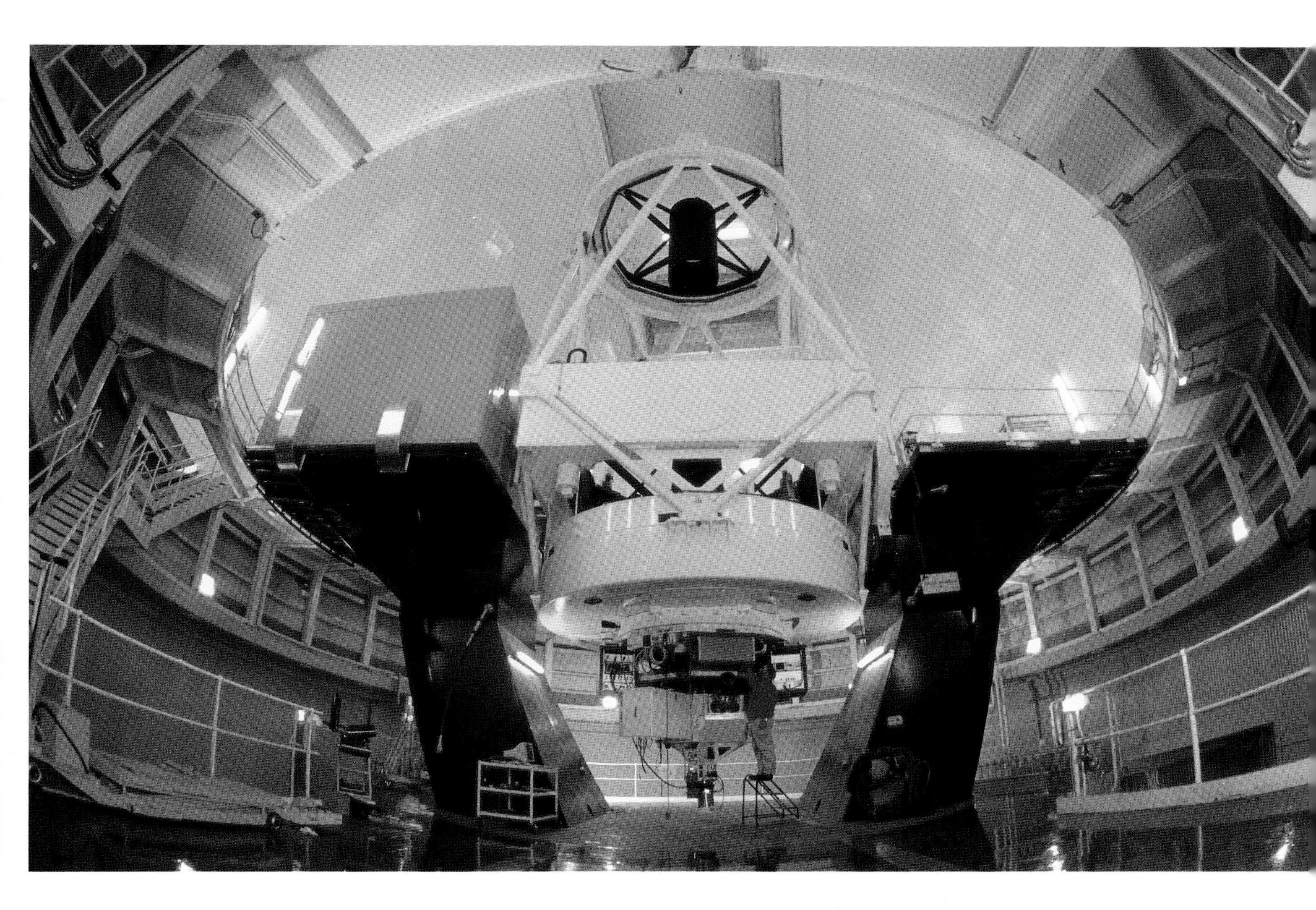

Roque de los Muchachos Observatory
La Palma, Canary Islands, Spain

The 4.2-meter William Herschel Telescope, named after the man who discovered Uranus, was completed in 1987. Featuring a lightweight alt-azimuth design with advanced, computerized motion control, it's located on a superior site at the lip of an extinct volcano, 2,400 meters (8,000 feet) above sea level. The Royal Greenwich Observatory of Great Britain operates the telescope in cooperation with Dutch and Spanish astronomers, making this the northern counterpart of ESO. Between these two observatories, European researchers have the entire sky under observation.

Cerro Tololo Inter-American Observatory (CTIO)
La Serena, Chile

CTIO is bathed in moonlight as astronomers travel from dome to dome using red flashlights to protect their night vision.

Operated by a consortium of 20 U.S. universities and supplemented with funding from the National Science Foundation, CTIO is the Southern Hemisphere observation point of NOAO, the National Optical Astronomy Observatories. NOAO also operates the Kitt Peak and Sacramento Peak Observatories in Arizona and New Mexico, and acts as a planning center for the development of large new American telescopes.

CTIO is situated in the same Chilean mountain range as ESO. Its largest telescope is a four-meter instrument completed in 1975.

Mauna Kea Summit
Hawaii, U.S.A.

(Previous pages, left, and above) Atop a 4,300-meter cinder cone on the island of Hawaii sits the highest major observatory in the world. On Mauna Kea the seeing is remarkably steady, and the night air is clear and dry most of the year. Several large telescopes have been built here, including the 3.6-meter Canada-France-Hawaii Telescope shown at left.

Many visitors find it hard to breathe at this altitude, and most discover that at first they can't think clearly. However, the exceptional viewing conditions are approached only by sites in Chile and the Canary Islands.

Many large telescopes are planned for Mauna Kea, and work is nearing completion on the huge 10-meter Keck Telescope—soon to be the world's largest. The above photograph was made by moonlight at the time that the Keck Observatory's foundation was being laid in the winter of 1987. My camera was positioned at the exact spot where the telescope was later mounted.

The Keck, owned and operated by the University of California and the California Institute of Technology, features a major advance in lightweight mirror technology. Instead of using a single, large 10-meter disk, the Keck employs 36 separate hexagonal reflectors. Computer controls will combine the light from each segment into a single, coherent image.

Hubble Space Telescope
Sunnyvale, California, U.S.A. ●

Orbiting telescopes like the Hubble promise to expand the boundaries of the observable universe by a factor of seven or more. Free from the effects of our shimmering, polluted, moist atmosphere, the totally unimpeded view from space means that even a relatively small 2.4-meter telescope like the Hubble can set amazing new records. For example, it will produce photos of Jupiter as sharp as those taken by Voyager (pages 160 and 166).

Hubble's designers calculate that it will see objects 50 times fainter, with an image clarity ten times sharper, than is possible through our largest earth-based telescopes.

Here at Lockheed Missiles and Space Company, where telescopic spy satellites are also built, the Hubble undergoes final check-out in a giant clean room.

Multiple Mirror Telescope (MMT)
Mt. Hopkins, Arizona, U.S.A. ●

Completed in 1979, this radical design collects light from six symmetrically mounted 1.8-meter telescopes. The combined aperture equals that of a single 4.5-meter telescope. Experience with the MMT has paved the way for future multiple-mirrored eyes on the universe.

Steward Observatory Mirror Lab
Tucson, Arizona, U.S.A. ●

Using the giant, twirling oven above his head, Roger Angel spin-casts prototypes for the huge eight-meter mirrors of the future. The spinning forces molten glass into a concave shape before cooling, bypassing the tedious task of grinding tons of glass out of flat mirror blanks. Here, Angel checks the curvature of a 1.8-meter disk prepared in this way.

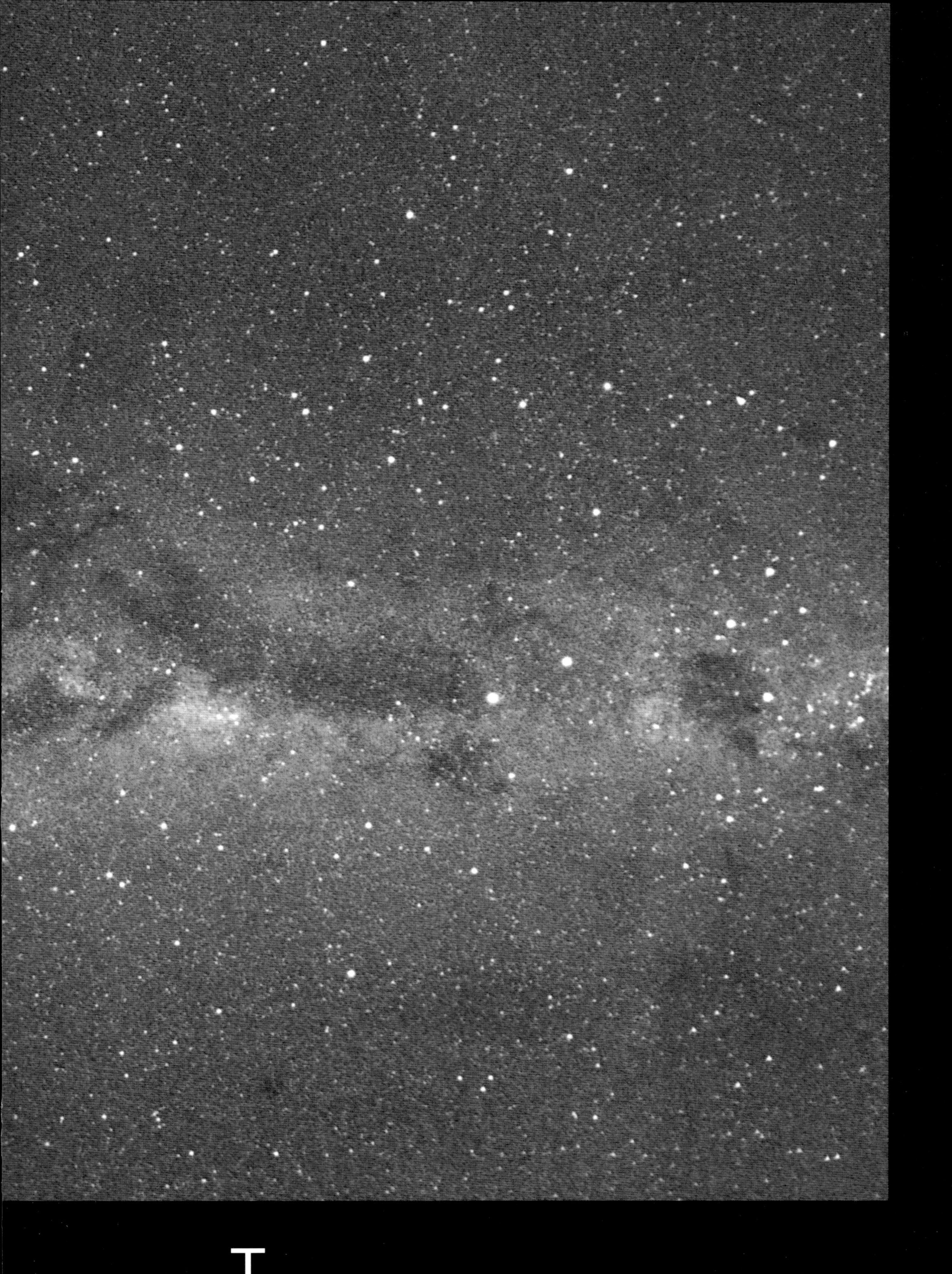

The diversity of the phenomena of nature is so great, and the treasures hidden in the heavens so rich, precisely in order that the human mind shall never be lacking in fresh nourishment. —Johannes Kepler

THE PROCESS OF DISCOVERY AND SUPERNOVA 1987A

Space research has advanced dramatically over the past century. In 1900, astronomers recorded their observations by sketching. Today, telescopes use sensitive, super-cooled electronic detectors to analyze light from space. Indeed, professional astronomers no longer even look through their telescopes. Instead, they sit in warm rooms and watch computer monitors.

All wavelengths of electromagnetic radiation, not just those of visible light, are probed with an array of new machines barely recognizable as "telescopes." Here, we'll see how such instruments recorded the most exciting astronomical event of the 1980s—the appearance of the brightest supernova (exploding star) seen from earth in 383 years.

On a cold, dark night in February, 1987, Ian Shelton, the University of Toronto's resident observer at Las Campanas Observatory in Chile, was tinkering in his off hours with a 25-centimeter (10-inch) telescope like those owned by hundreds of thousands of amateurs. After winding its primitive clock drive, Shelton used the scope to make an ordinary black and white photograph of the galaxy nearest the Milky Way (see front cover). An hour later, in his darkroom, he noticed a curious speck on the negative. Thinking at first that it was dirt, he took another look outside, and thus became the first to identify a brilliant "new" star, Supernova 1987A, located in a galaxy called the Large Magellanic Cloud (LMC). The spectacle would rock the astronomical world by providing a bright, easy-to-study view of the forces that create the building blocks for planets and life in the universe. The last "naked eye" supernova had appeared nearly four centuries earlier, in 1604, five years before Galileo first turned a telescope toward the heavens.

Supernova 1987A was a stellar goldstrike for astronomers. The star has been studied in detail by almost every type of telescope in existence—not just optical telescopes, but also neutrino detectors, infrared telescopes, ultraviolet telescopes, X-ray telescopes, gamma-ray telescopes, and radio telescopes operating at every wavelength.

Exploding stars represent the most significant stage of stellar evolution, for they give birth to all the heavy elements in the universe. Before exploding, a star "burns" furiously, exhausting every possible fusion reaction among the elements in its core, and in the process generating an array of heavier elements including carbon, oxygen, silicon, gold, iron, and lead. When no more elements can be fused into existence, the star collapses under its own weight, triggering a momentous explosion which, for a few brief moments, releases as much energy as the combined output of all the billions of stars in the galaxy.

Hurled into space in billowing clouds of stardust, the elements eventually condense into the objects we call planets. The explosion's shock wave also triggers the collapse of nearby gas clouds, creating new stars. No wonder astronomers were excited! Supernova 1987A presented a perfect opportunity to study the chemical evolution of the universe.

Solar astronomers reassure us that we needn't fear a supernova explosion of our own star. Luckily for us, the sun burns too slowly. But even so, it holds an important secret: how to produce a virtually unlimited supply of energy, safely and steadily.

By solving the mysteries of solar fusion, astronomers may provide a permanent answer to earth's energy needs. Therefore, we'll follow our look at supernova research with a peek at solar astronomy.

Las Campanas Observatory
La Serena, Chile

After discovering Supernova 1987A, Ian Shelton was asked to take nightly measurements of its brightness with this 61-centimeter telescope. The exploding star is the upper of the two red spots visible above the telescope in this photo, just below the fuzzy area that is the LMC galaxy. The streak of light below and to the right of the LMC is a meteor ... a small piece of star dust burning up as it enters the earth's atmosphere.

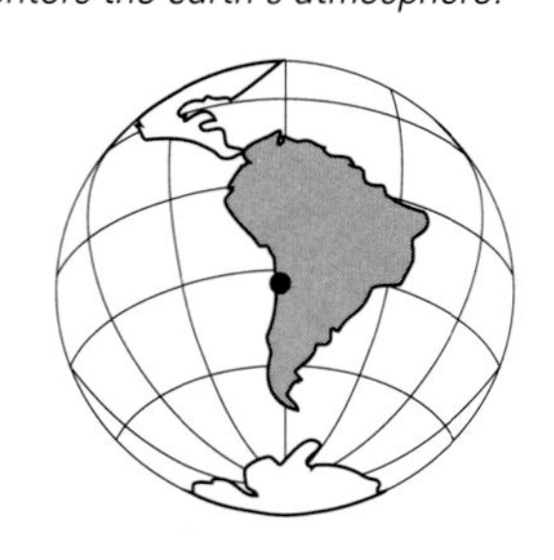

Photo by Roger Ressmeyer with Ian Shelton

Amateur Astronomy

Not restricted to specific tasks, amateur astronomers are in some ways better equipped to spot new comets and extragalactic supernovae than their professional counterparts. And when an amateur does find a new comet, it bears his or her name. "For a while, I was a mile high," says Doug Berger, one of the three co-discoverers of Comet Kobayashi-Berger-Milon.

(Top left) Hobbyist Mary Engle adjusts the cradle that holds her homemade 41-centimeter telescope mirror at a workshop in California's Chabot Science Center. She is performing a precise optical test on the mirror's polished surface, the results of which will reveal imperfections as small as 25 angstroms (one-tenth of a millionth of an inch). To do this, she places a tiny grating made up of perfectly straight, illuminated lines at the mirror's center of curvature. The glowing lines, when reflected by the mirror, are seen as curved bands in the photograph. The test graphically reveals a dimple-like depression in the mirror's center. Before rubbing the final curve into her mirror, Engle will remove the depression with additional polishing.

(Bottom left) When this picture was taken in 1985, Don Machholz of California had discovered two new comets that now bear his name. By 1990, his total was four. He's spent thousands of hours with this telescope in the Santa Cruz Mountains, looking for these dim, fuzzy spots in the sky. The dark eye patch screens distracting light from his unused eye.

(Above) Silver-haired John Dobson is the Pied Piper of amateur astronomers. He perfected a simplified telescope mounting that has enabled astronomy buffs to afford much larger telescopes, ideal for viewing dim, "deep-sky" objects like galaxies, nebulae, and star clusters.

The "Dobsonian" mount is seen here supporting a 46-centimeter scope that Dobson calls "The Little One." It rotates the telescope on simple alt-azimuth bearings like those on the Soviet BAT. Years ago, Dobson founded the San Francisco Sidewalk Astronomers, a volunteer group dedicated to spreading knowledge about the heavens.

(Above) Jack Marling (foreground) has inspired amateur astronomers to take photographs rivalling those of the big observatories in beauty. Marling left a job at Lawrence Livermore National Laboratory to found Lumicon, a company dedicated to supplying the telescopes and accessories needed to make perfectly guided time exposures of dim celestial objects.

(Top right) By the time this photograph was made in 1987, the Rev. Robert Evans had discovered 15 supernovae in distant galaxies—a record. From his suburban neighborhood in Hazelbrook, Australia, Evans spends moonless nights repeatedly scanning several hundred galaxies, looking for bright spots. Supernovae can appear to be far brighter than entire galaxies consisting of billions of stars.

(Bottom right) In 1986 telescope designer Mike Simmons tested a new invention at Ayers Rock in Australia. His device fits many off-the-shelf amateur telescopes, converting them into computer-controlled platforms that can find thousands of dim space objects at the press of a button. In the course of this shoot, he observed 600 clusters and galaxies in a mere 32 minutes. His invention also helps avoid false sightings. Comet hunters are warned by a beep and a message display when they come within eyeshot of a nebula that might be mistaken for a comet.

European Southern Observatory (ESO) La Silla, Chile

Supernova 1987A exploded far south of the celestial equator, and appears high in the sky only from the Southern Hemisphere. The far southern sky contains many jewels: the bright stars in the Southern Cross, the luminous Magellanic Clouds, and dozens of exceptionally beautiful nebulae, star clusters, and galaxies. So when astronomers like Patrice Bouchet (above) are assigned to the European Southern Observatory, it's a thrill.

Bouchet directed ESO's infrared research into Supernova 1987A using a one-meter telescope in the dome just behind him. So bright was the supernova that some of the best observations with this telescope could be made during daylight hours.

(Right) Its reflective surface gleaming in the moonlight, ESO's new submillimeter telescope looks for short-wavelength radio emissions from the supernova's shock wave. The 3.6-meter telescope in the background (also page 54) studied the explosion as well, using an infrared "speckle" camera linked to a computer which analyzed and combined thousands of electronic images. The camera/computer combination strips away the distorting effects of the earth's atmosphere, producing pictures of extremely high resolution.

Cerro Tololo Inter-American Observatory (CTIO)
La Serena, Chile

Biggest isn't necessarily best. Supernova 1987A prompted astronomers to reactivate the smallest telescope at CTIO. The 41-centimeter scope is ideal for closer and brighter phenomona, but it still needed "telescopic sunglasses" to reduce the star's glare to measurable levels. At sunset, Chilean research assistant Mario Hamuy prepares to attach a shielding screen; the small holes reduce the telescope's effective diameter to a mere eight centimeters. Night after night Hamuy used this telescope to measure the supernova's "light curve" in ultraviolet, red, and infrared wavelengths.

In a cage suspended beneath the mirror of the largest CTIO telescope, Harvard astronomer Costas Papaliolios attaches an American speckle camera with the help of graduate students Jim Beletic and Bruce Sams. The high-resolution, computerized imaging machine used all of the light focused by the four-meter telescope to produce a close-up picture of the region around the exploding star.

One such image revealed a brilliant, short-lived "mystery spot," two light-weeks away from the explosion, that was almost a tenth as bright as the supernova itself. One astronomer joked that the flash of light was the exhaust from an accelerating star ship evacuating aliens from the supernova's inferno.

Kuiper Airborne Observatory
Christchurch, New Zealand

NASA mobilized several research branches to study the supernova. NASA Ames at Moffett Field, California deployed a flying observatory with a 91-centimeter infrared telescope. Based out of temporary headquarters in the city of Christchurch, missions were flown with a rotating staff of visiting American astronomers.

From an altitude of 12,500 meters (41,000 feet), high above most of the infrared-absorbing water vapor of the lower atmosphere, astronomers detected emissions of nickel, cobalt, and iron amid the remnants of the celestial blast. Above, the Kuiper's telescope peers out through a sliding hatch in the Lockheed C-141 Starlifter cargo plane.

At right, Mission Director Carl Gillespie (seated, foreground) serves as the link between these scientists, their apparatus, and NASA's flight crew during an all-night mission circling high above the South Pacific.

CSIRO Radio Observatory
Parkes, Australia ●

Seen here as it revolves during maintenance work, the 64-meter radio dish at Parkes was also used to detect emissions from the supernova. Scientists theorize that these radio signals were created by the impact of the explosion's shock wave against the gaseous "stellar wind" that encircled the supernova's progenitor star.

The telescope was completed in 1961, and is equipped to record radio waves from 5 to 2,000 millimeters in length. Astronomer Dick Manchester arranged with colleague David Jauncey to link this dish by microwaves with another at the Tidbinbilla Space Tracking Station, 275 kilometers away. The combined dishes create an "interferometer" capable of "resolving"—or revealing the detail of—the supernova's shock wave.

Molonglo Observatory Synthesis
Telescope (MOST)
Canberra, Australia ●

(Following pages) This was the first radio telescope to detect the supernova. When news of the discovery reached him through a grapevine of amateur astronomers, officer-in-charge Duncan Campbell-Wilson junked his usual observing routine and directed his machine toward the cataclysmic burst.

Here, Campbell-Wilson stands at one end of the mile-long "ear" which served as a prototype for other advanced radio telescopes like the Very Large Array in New Mexico (page 172). MOST takes 12 hours to make a complete exposure.

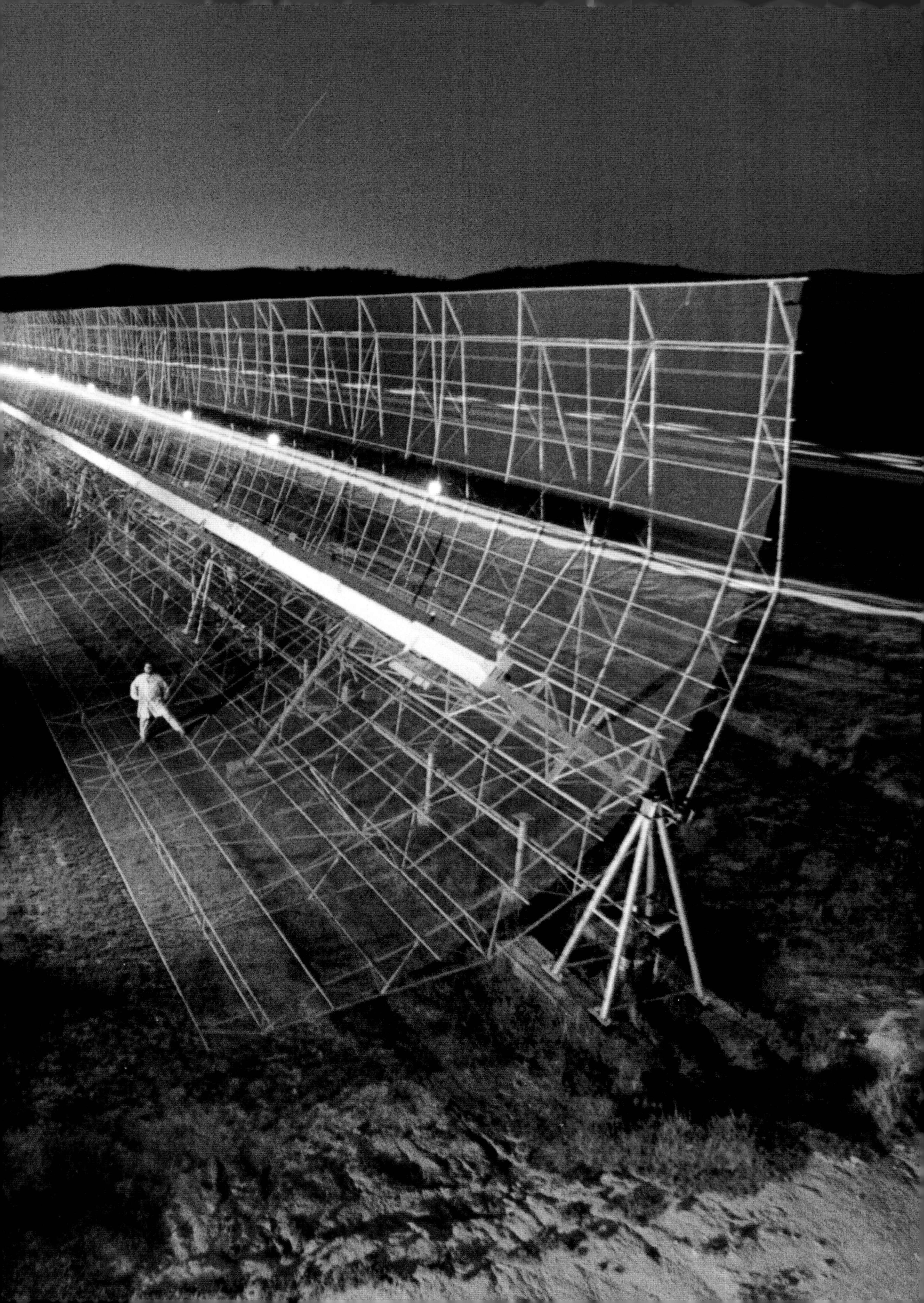

Woomera Range
Woomera, Australia

Using this two-stage rocket, NASA launches an X-ray telescope to an altitude of 300 kilometers in an effort to detect X-rays from the supernova's core. The high break in the trail is where the first and second stages separate, just four seconds into the flight. The dimmer trail to the right of the bright streak records the fall of the first stage eight kilometers back to earth.

A few hours later, at sunrise in the red Australian desert, the recovery crew disconnects the payload's parachute before early morning winds begin blowing. A gust could send the telescope flying across the desert like a loose umbrella. The payload, prepared by the astronomy department at Pennsylvania State University, will be reused on future flights.

Small rockets like this one provide a quick, inexpensive way of getting sensors aloft, but their telescopes have to work fast because they are above the atmosphere for only a few minutes before falling back to earth. Despite that limitation, these sounding rockets are perfectly suited to the study of this supernova. NASA has launched over 2,500 of them in the past 30 years.

NASA Balloon Launch Facility
Alice Springs, Australia

Sending a 1,400-kilogram balloon payload to an altitude of over 40 kilometers requires skill and a bit of luck. Above, a helium-filled balloon lifts off behind a truck that carries a "Gamma Ray Imaging Payload" (GRIP) from the California Institute of Technology. When the balloon reaches the end of its tether, it must be aligned directly over the payload truck or the instrument will crash against its supports. The truck's driver is prepared to speed off in any direction to assure this alignment at the moment of release. At high altitudes where the atmospheric pressure is lower, the polyethylene balloon expands to over 180 meters in diameter.

On this flight, the gamma ray telescope found evidence of radioactive cobalt in the supernova's core. To record such findings, scientists used off-the-shelf Sony Beta VCRs. The video recorders' enormous data capacity makes them ideal for this application. NASA's principal balloon launch facility is at Palestine, Texas, an area far from commercial airline routes.

NOVA Laser
Livermore, California, U.S.A. •

Even as the supernova provided astronomers the opportunity to study nuclear reactions on a cosmic scale, increasingly accurate simulations of both stellar fusion and supernova bursts were taking place in laboratories and supercomputers around the world.

At left, nuclear fusion is triggered in a vacuum chamber at Lawrence Livermore National Laboratory. Scientists bombard a fusion pellet—a one-millimeter glass sphere containing deuterium and tritium—with the combined blast of ten brilliant, blue laser beams. In just one-billionth of a second, this, the world's largest and most powerful laser, can generate temperatures found at the sun's core. The light is so bright that my cameras had to be shielded with ultra-dense filters.

Fusion energy fuels are abundant on both the earth and the moon, and could last humanity for millions of years. Even today, scientists foresee fusion power plants that will produce only short-lived radioactivity as a by-product. It is conceivable that such plants could also be used to transform the dangerous long-lived radioactive waste from today's fission plants into safer, ephemeral by-products that can be recycled after 50 to 100 years.

NASA Goddard Space Flight Center
Greenbelt, Maryland, U.S.A. •

A satellite orbiting 35,000 kilometers above the Atlantic Ocean recorded this ultraviolet spectrum of Supernova 1987A, sixteen hours after Ian Shelton's discovery. The satellite, called the International Ultraviolet Explorer (IUE), was used by Harvard astrophysicist Robert Kirshner (right) and NASA scientist George Sonneborn to identify the star that had exploded.

In ultraviolet wavelengths, neighboring stars appear brighter than the supernova's burst. By comparing pre-explosion pictures of the same star field, the supernova's progenitor star was identified four months after the explosion was seen from earth.

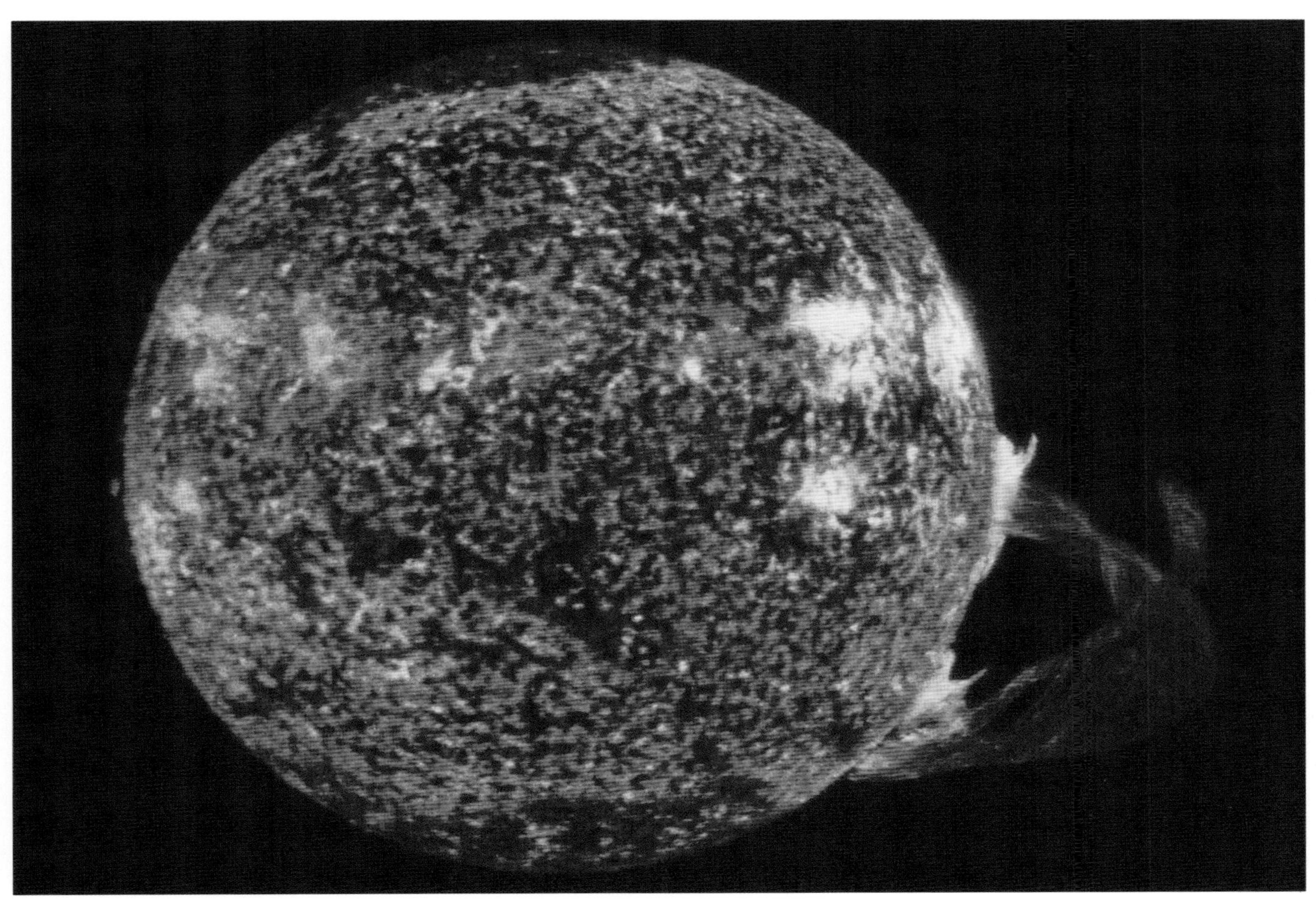

Photo by Skylab – Naval Research Lab / NASA

McMath Solar Telescope
Kitt Peak, Arizona, U.S.A.

The world's largest sun telescope produces a solar image 76 centimeters in diameter. Astronomer Emeritus Keith Pierce, who designed the optics, wears protective sunglasses while measuring sunspots in the image. During periods of peak sunspot activity, the sun's surface is dotted with storms, spots, and eruptions.

The McMath is used 24 hours a day, because light from its 1.5-meter mirror is bright enough to allow the study of stars besides our own. Pierce remembers a day in the late 1960s when Apollo astronauts gathered around this same table to study the moon's image for prospective landing sites.

Solar Flare
photographed from Skylab
Low Earth Orbit

When the moon-landing program was completed, NASA flew one last Saturn V rocket before retiring the giant booster. In May, 1973 it lofted the Skylab space station, a cylinder 36 meters long that would be inhabited by three crews of astronauts on successive missions of 28, 59, and 84 days.

Skylab carried numerous astronomical instruments, including an ultraviolet slitless spectroheliograph that made spectacular pictures of the sun.

This image (above) shows a violent solar flare nearly 500,000 kilometers high that sent charged particles flying outward into the solar system. Study of the sun, the star nearest to us, goes hand in hand with the study of more distant stars and supernova.

Aurora photographed from Space Shuttle Challenger Low Earth Orbit

Charged particles hurled into space by tumultuous solar flares (previous page) can severely disrupt radio and electrical transmissions on the earth's surface. When the eruption points in the direction of earth, particles arrive here in a few hours and interact with the donut-shaped magnetic fields that surround the planet. The charged particles are pushed north and south toward the poles, where they finally tumble earthward.

When they strike the upper atmosphere, the particles add energy to atoms in the air. The extra energy is released in the form of a luminous, shimmering stream of light called the aurora borealis, or northern lights. South of the equator, the same phenomenon is known as the southern lights (aurora australis). The display here was photographed from orbit in spring, 1985 by a NASA astronaut using high-speed film.

Photo by Robert Overmyer – NASA

Solar Eclipse photographed from Jet Plane North of Hawaii, U.S.A.

Total solar eclipses are particularly intriguing to solar astronomers. By some serendipitous cosmic accident, the moon and the sun have almost identical angular diameters in the sky, as viewed from earth. When the moon is aligned directly between us and the sun, it shields the blinding light of the sun's surface but leaves the pearly white corona visible on all sides of the moon's disk.

The best place to view the eclipse shown at left was from a jet plane that intersected the moon's shadow for a few brief moments over the Pacific. Aside from the pictures taken by David Malin with observatory telescopes, this is the only "composite" photo in the book. Because film is more sensitive to contrast than is the human eye, my pictures didn't reflect what my eyes had seen. In order to more accurately represent the actual experience, I used a double exposure to combine two pictures in the darkroom. One photo showed the sun's bright "diamond ring" at the end of totality. The other showed the same scene, but with a longer exposure that revealed the moon's shadow striking the clouds.

Above, solar astronomer Jay Pasachoff prepares to view the partially-eclipsed sun from the window of our jet. His wife and daughters don solar filters to protect their eyes. During totality, it is safe to look directly at the sun's corona. As the event ends, however, and the sun's rays burst out from behind the moon, the eyes must be closed or turned away to prevent blindness. It's an especially vulnerable moment because the eyes have adapted to the darkness of totality, and the pupils are wide open.

Crimean Astrophysical Observatory
Simeiz, U.S.S.R.

During a total solar eclipse, the moon's shadow sweeps across the earth at about 1,900 kilometers per hour. Anywhere within the slender path of that shadow, the entire face of the sun will be briefly blocked out. Depending on the geometry of the eclipse, totality can last just an instant, or as long as seven minutes.

On average, total eclipses occur just every 18 months, and even then the path of totality only covers a small area on the earth's surface. So scientists have built telescopes like this one, called coronagraphs, to simulate total eclipses. A metal disk is used to obscure the sun, resulting in an eclipse-like view of the bright, inner section of the sun's corona. The Soviet Union's program has eight such instruments.

Solar Eclipse Expedition
March 7, 1970
Miahuatlán, Mexico

This picture was made on my first eclipse expedition, at age 16, at a site where totality lasted three and a half minutes. It shows the "diamond ring" phenomenon which occurs for a few seconds at the beginning and end of the event. A tiny part of the sun's face peeking out from behind the moon creates a "diamond" for the coronal ring.

Total eclipses offer one of nature's most spectacular displays. During totality, the stars come out, the wind blows up, the temperature drops, and animals make odd noises as their schedules are disrupted by untimely twilight. At the end, the sky begins to brighten as patches of light and dark, called shadow bands, race along the ground. As quickly as it begins, the spectacle ends. It's worth the trip.

Photo by Apollo 4 – NASA

Our sun is one of 100 billion stars in our galaxy. Our galaxy is one of billions of galaxies populating the universe. It would be the height of presumption to think that we are the only living things in that enormous immensity. —Wernher von Braun

Are we alone? Considering the millions of trillions of stars out there, shouldn't there be many thousands, or even millions, of earth-like planets? And if life exists in such abundance here on earth, why not elsewhere in the universe?

In the 1950s, the famous Miller-Urey experiment suggested that we are probably not alone among the stars. Stanley Miller had simulated one possible version of the lifeless environment on a primordial earth, and he found that there is a natural process which generates organic molecules out of inorganic gases. Although the end product was not life itself, the experiment showed how natural processes can favor the development of life without any special intervention by a god or a cosmic force.

If life exists on other planets, how likely are intelligence and advanced technology? In light of our own dreadful failure (so far) to protect planet earth, is it conceivable that any life form using technology could survive for more than a few centuries? We seem discouragingly close to destroying our environment with toxic waste, global warming, ozone depletion, and simple exhaustion of precious resources. But what if an advanced extraterrestrial civilization had managed to flourish and endure for millions of years? Could we discover its secrets of survival? Could we follow its example to preserve our own planet? Could we learn how to keep the earth pristine while still enjoying the technology that has possibly become our final addiction?

Finally, wouldn't contact with extraterrestrials make the differences between races and cultures on earth seem trivial? Wouldn't that awareness pave the way for global peace? And what impact might interstellar communications have on the religions of our world?

Exobiology, the study of life beyond earth, raises many such intriguing questions. And some believe that we are on the brink of discovering extraordinary answers. While communicating with life elsewhere in the universe, we may learn to better understand and protect humanity. To visit space places at the forefront of this research is to gain a new appreciation for the meaning of life—where science and metaphysics are one.

University of California, San Diego
La Jolla, California, U.S.A.

Stanley Miller uses his original apparatus to re-create the famous Miller-Urey experiment of the early 1950s. Gases in the glass bulb—hydrogen, methane, ammonia, and water vapor—simulate the earth's atmosphere shortly after its creation. The incandescent electrodes imitate a miniature bolt of "lightning." The result is a tar-like residue containing amino acids, the basic building blocks of proteins and DNA.

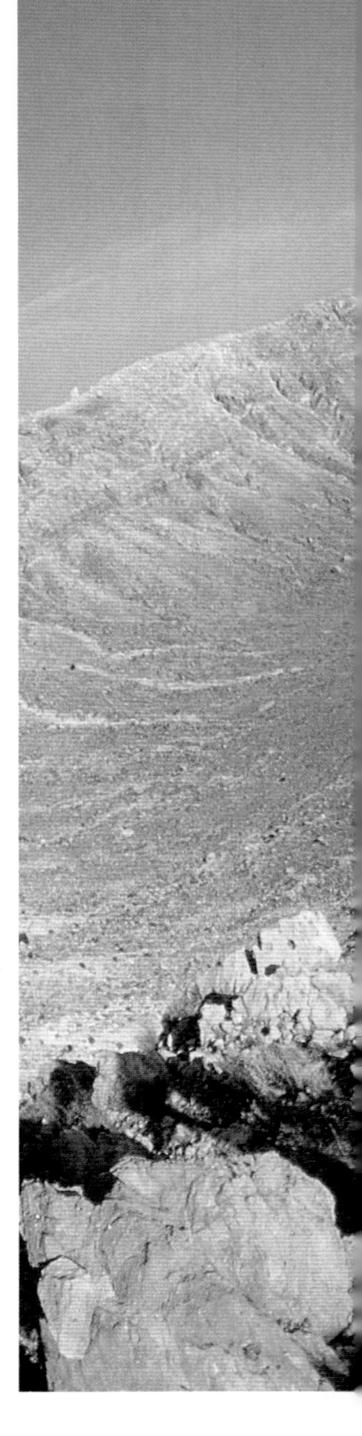

Lawrence Berkeley Laboratory
Berkeley, California, U.S.A. •

Stumbling upon a possible explanation for the extinction of the dinosaurs, father-and-son team Walter and Luis Alvarez (above) discovered abnormally high quantities of the rare element iridium in a 65-million-year-old layer of sediment found around the globe.

They theorized that iridium and other rare elements in the sediment were distributed by the explosive impact of a comet or asteroid, the residue of which blotted out the sun with a worldwide cloud of atmospheric smoke and debris. In a matter of weeks temperatures dropped, photosynthesis of plankton and plants was arrested, and the fragile food chain was obliterated. Nearly half of the life forms on earth became extinct. Thinking about that makes it easy to appreciate the fragility of the environment, and our dependency on its good health.

Meteor Crater
Flagstaff, Arizona, U.S.A. •

Planetary geologist Eugene Shoemaker did his doctoral dissertation on Arizona's giant impact crater. The crash occured some 22,000 years ago, when a nickel-iron meteor some 30 meters across plowed into the earth at a speed of about 16 kilometers per second and blasted a hole nearly 1.6 kilometers wide. Shoemaker and his wife Carol now search the skies for "earth-crossing" asteroids. A collision with any one of them in the near future is unlikely but would be catastrophic. Some scientists suggest we should prepare a rocket that could intercept a menacing asteroid and nudge it into a new orbit around the sun that would miss the earth.

Purple Mountain Observatory
Nanjing, China ●

Like a gargantuan pair of binoculars, this "twin astrograph" is used to plot the movements of asteroids and comets that might threaten earth. Since the instrument makes two separate photographic plates at the same time, there can be immediate confirmation when a new object is sighted.

Beijing Astronomical Observatory
Xinglong Station, China ●

The 60-centimeter Schmidt telescope above is used as a camera to track minor planets—either with wide-angle photographic plates or with modern charged-couple devices (CCD's), which record the image electronically.

Schmidt cameras are admirably suited for the wide-field surveys used to discover new asteroids and comets, because they compress a vast expanse of sky into a single bright image. Each has a thin, transparent correcting lens at the front of the tube, a larger concave mirror at the back end, and a curved focal plane suspended in the middle, where the film goes.

Important northern sky surveys have been made with the 122-centimeter (48-inch) Schmidt at Palomar. Eugene and Carol Shoemaker use a second, smaller, 46-centimeter Schmidt at Palomar to search for earth-crossing asteroids. The most popular amateur telescope for serious viewing is the 20-centimeter (8-inch) Schmidt-Cassegrain design. It uses a second mirror to project the image outside of the tube, where it can be viewed with an eyepiece. Over 100,000 of these compact telescopes have been sold since 1970, at a cost of over $1,000 each.

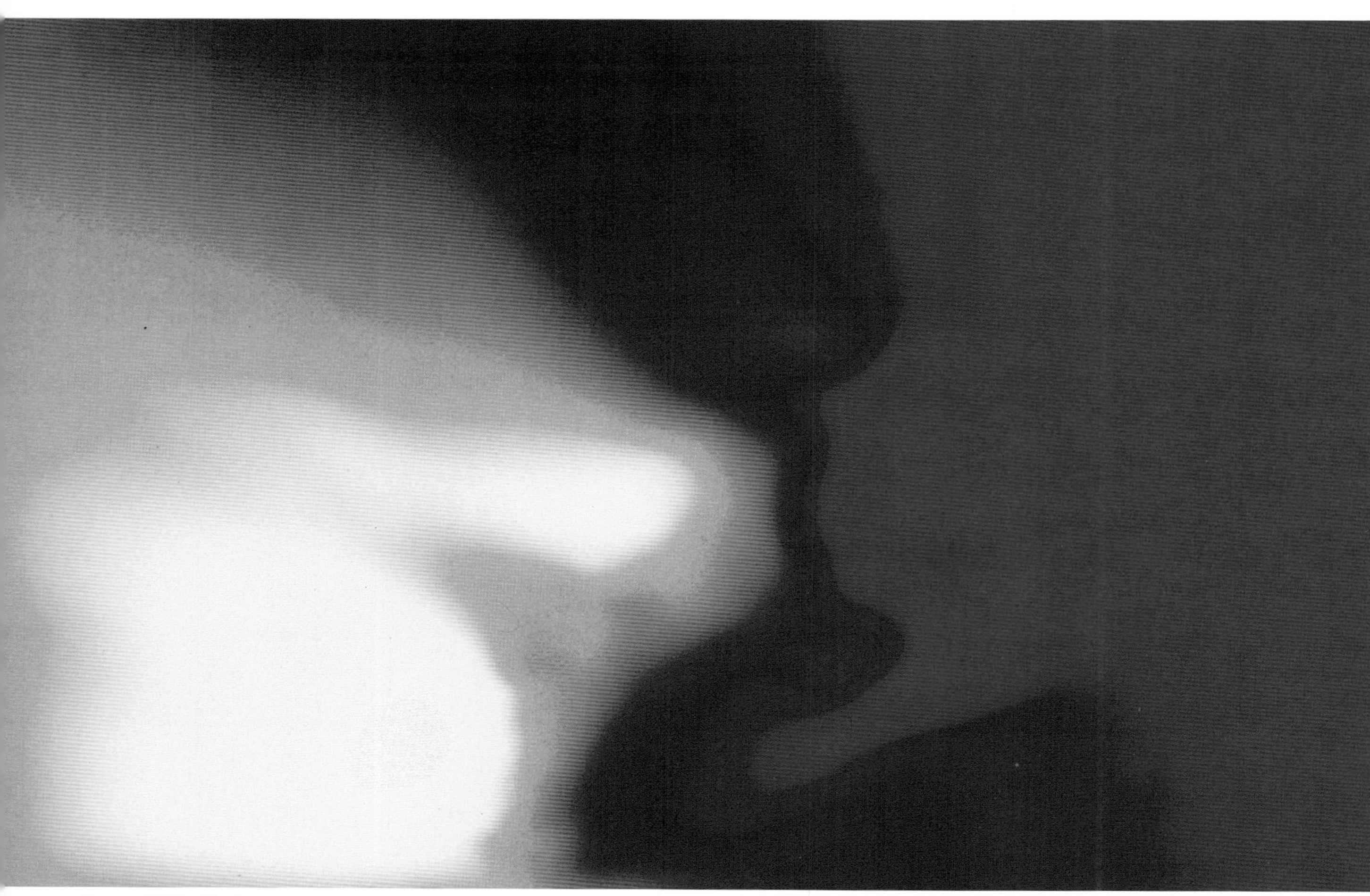

Photo ©1986 Max-Planck-Institut für Aeronomie, courtesy of Alan Delamere & Harold Reitsma, Ball Aerospace Systems Group

Halley's Comet Nucleus photographed by Giotto Probe

Organic molecules have been detected in comets, leading scientists to suppose that life might arise in a wide variety of situations. While on a kamikaze mission into the heart of Halley's Comet in 1986, the European Space Agency's Giotto probe made this close-up picture of the dirty ice ball found at the comet's core. A fraction of a second after this picture was made, the camera was damaged by a collision with cometary debris streaming off Halley's 16-kilometer-long nucleus.

The evolution of life on earth may have been accelerated by the impact of comets. It's possible they seeded the planet with amino acids and perhaps even complex organic molecules from the outer fringes of the solar system. Analysis of the light given off by comets reveals the presence of hydrocarbons and other organic materials. Organic molecules have also been discovered inside meteorites—pieces of rock that have crashed through our atmosphere and landed on earth. Even in interstellar gas and dust clouds, organic material has been detected.

CSIRO Radio Observatory Parkes, Australia

This 64-meter radio telescope was modified during the Giotto mission to make it a receiver for pictures and data from the space probe.

In the eight days before Giotto's encounter with Halley, one Japanese and two Soviet probes also explored the comet up close. Precise positional data from these earlier passes was shared with the Giotto scientists, enabling them to fine-tune Giotto's trajectory. As a result of this international cooperation, Giotto was able to make an even closer approach to the nucleus than did its predecessors—a mere 605 kilometers.

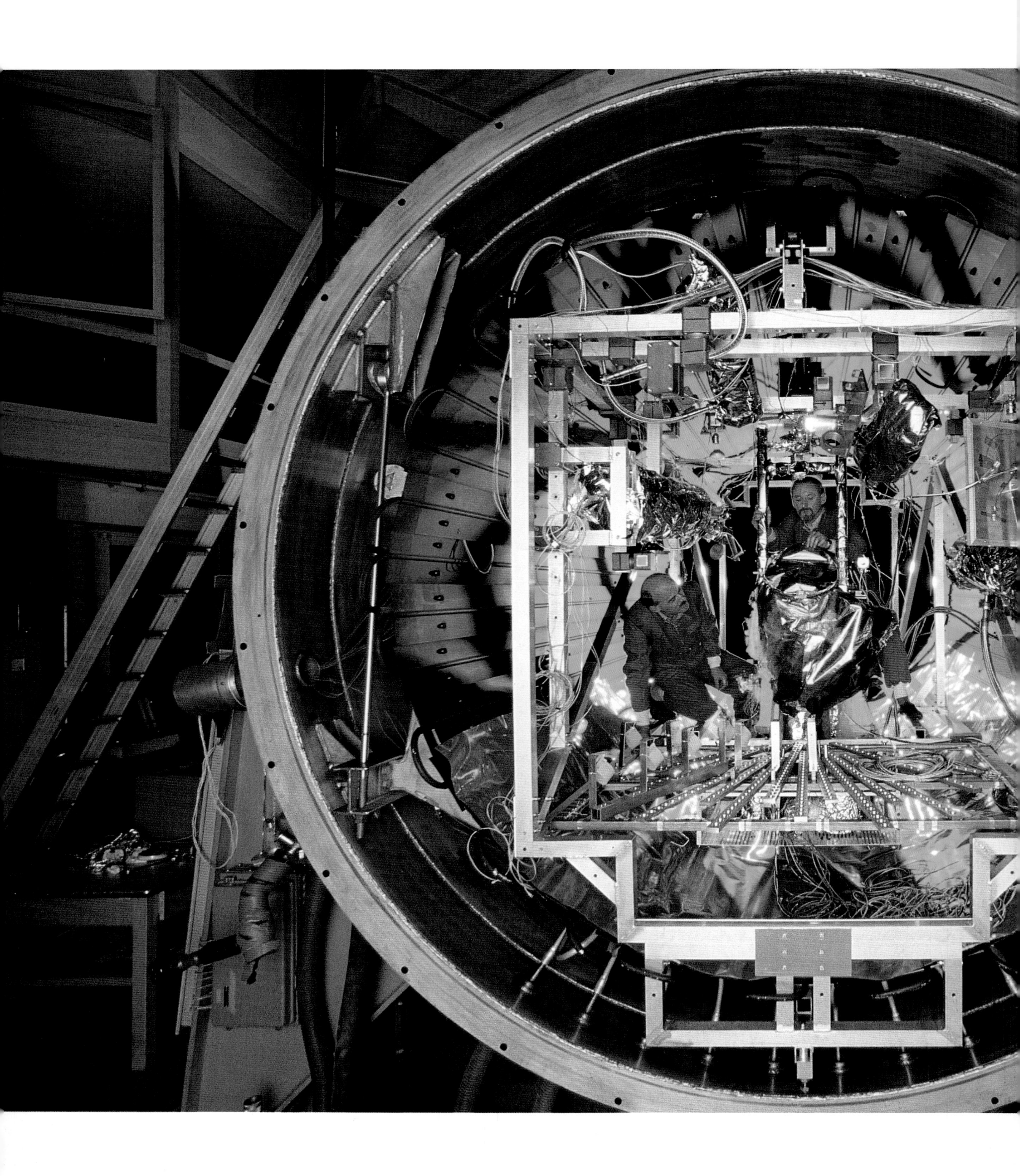

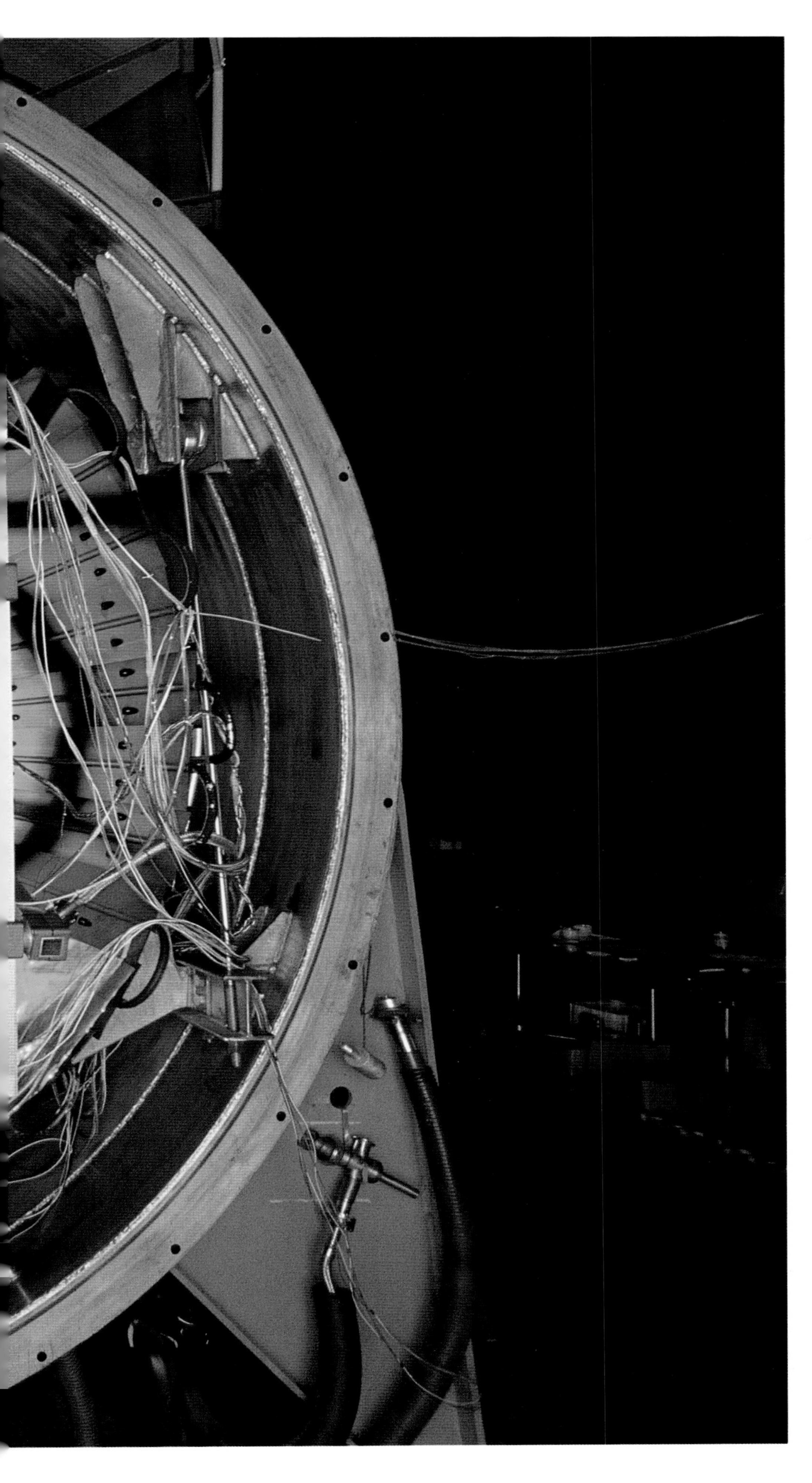

**DLR Headquarters
Cologne, Federal Republic
of Germany**

A number of German scientists have become preoccupied with comets in recent years. Evidence that they carry the chemical building blocks of life is but one reason; comets also provide clues to the origins and early history of the solar system.

Plans are now under way to land a probe on a comet's nucleus, take samples, and return them to earth. In researching the parameters of this tricky mission, scientists at DLR, the German Aerospace Research Establishment, simulate a comet's nucleus and its environment in space.

Just minutes before sealing this vacuum chamber, they prepare the artificial comet held in the central bucket by adding liquid nitrogen. When the air has been evacuated, the sample comet is bombarded with intense light that simulates sunlight in the inner solar system. Sensors in the chamber record the resulting jet of gas and particles from the "nucleus." The information gleaned about the comet's surface strength, texture, and accompanying debris cloud will enable engineers to design the comet lander.

Because the probe must exactly match the path and speed of its target, the actual comet mission will involve a complex trajectory. This will be far more difficult than just intersecting a comet for a few moments, as did the 1986 probes to Halley.

The orbits of comets are far more elongated than the orbits of planets. How did they get that way? The theory goes that comets normally circle the sun in a giant cloud located in the far reaches of the solar system, way outside the most distant planets. Called the Oort Cloud, it is made up of debris left over from the creation of the sun, planets, and asteroids. The cloud might be as large as two light-years in diameter. When perturbed by the gravitational force of a passing star, which happens about once every million years, some of its comets will be knocked into the inner solar system. If they pass near a massive planet like Jupiter, their paths can be further perturbed into elliptical orbits that pass near the sun on a regular basis, like Comet Halley.

NASA Johnson Space Center
Houston, Texas, U.S.A. •

Situated next door to a facility that stores many of the Apollo moon rocks, NASA's Cosmic Dust Lab studies the remnants of comets' tails. These microscopic particles have been plucked out of the upper atmosphere by high-flying planes outfitted with sticky collectors, a sort of hi-tech fly paper.

As a comet's nucleus speeds through the inner solar system, the sun vaporizes some of its icy core, giving the comet the appearance of an interplanetary Roman candle. Dust and gas accumulated during millions of years of comet fly-bys forms a cloud around the sun, visible before sunrise and after sunset at extremely dark locations as a dim glow called the "Zodiacal Light."

The earth's atmosphere sweeps through this cloud of dust, gathering comet particles that gradually settle to the ground. But when the comet dust is collected in the upper atmosphere, it can be separated from terrestrial dust.

Once in the Cosmic Dust Lab, the particles are stored in super-clean glove boxes like this one. The pressure inside the boxes is kept higher than the surrounding air so that any leaks will occur outward rather than in, keeping the samples clean. To reach into the box, you put your hands and arms into the inflated rubber gloves, turning them outside-in.

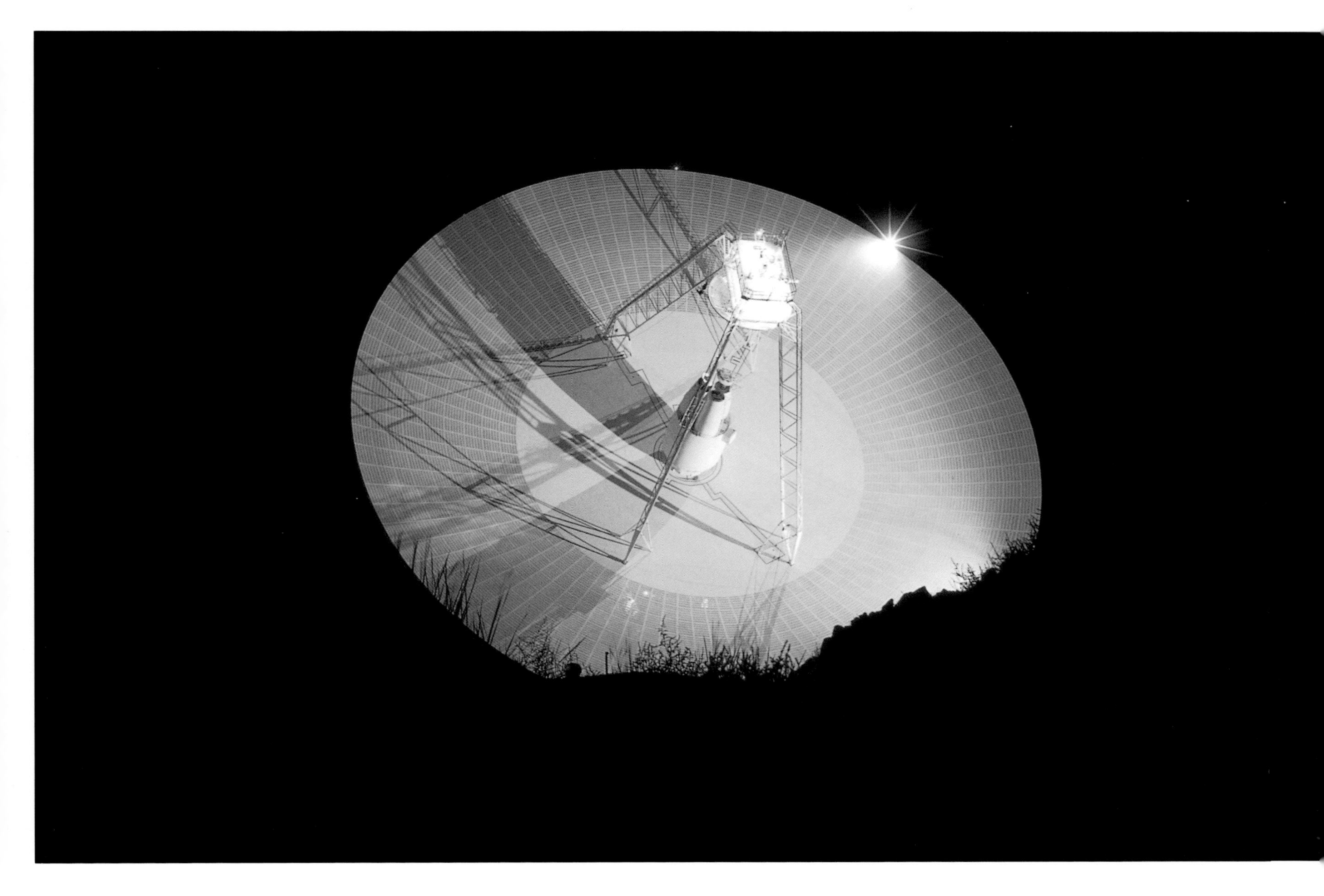

NASA Deep Space Network (DSN)
Goldstone, California, U.S.A. ●

With new computer technologies, the time has come for SETI, the Search for Extra-Terrestrial Intelligence. The Deep Space Network, built to communicate with our planetary probes and moon missions, is being joined by other radio telescopes in a globally coordinated effort to listen for technologically created signals from space.

NASA is building a data analysis system for this project that will conduct a search 10 billion times more comprehensive than all previous radio searches combined. At its heart is a device called a spectrum analyzer, which can search millions of microwave radio frequencies simultaneously.

One thousand nearby sun-like stars will be targeted for careful study, while the entire sky will be scanned in lesser detail. The project should be completed by the year 2,000.

Should SETI succeed in identifying intelligent signals, they might come from an extraterrestrial version of "I Love Lucy" that simply leaked away from a distant planet. Or, some scientists speculate, we might tap into an interstellar radio network that already links many alien civilizations.

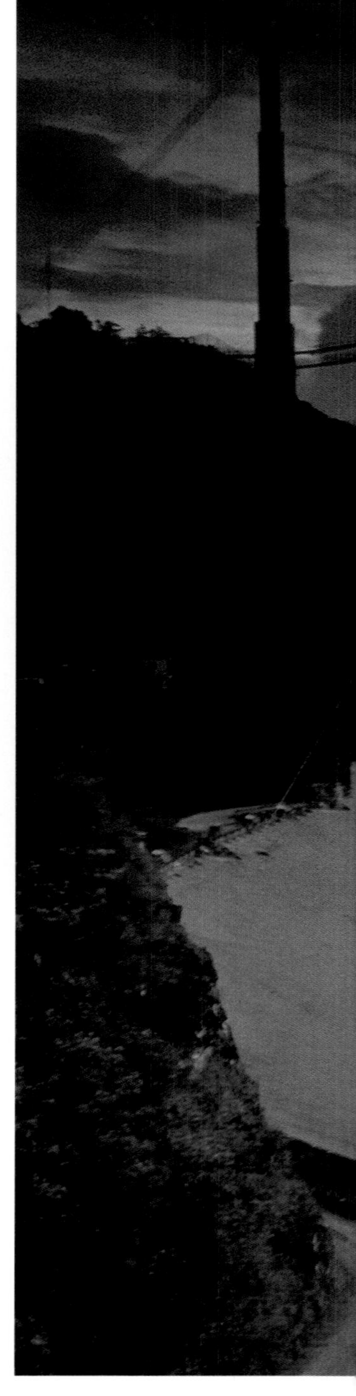

The Arecibo Observatory
Arecibo, Puerto Rico, U.S.A.

The world's largest radio telescope, 305 meters (1,000 feet) in diameter, has joined SETI. The aluminum wire and mesh dish has created its own ecosystem beneath its surface, where only partial sunlight reaches the ground.

Since the dish itself is too large to be pointed, the telescope is "steered" by the earth's rotation and by a movable receiver suspended in the dish's focal plane.

In 1974 this telescope transmitted the "Arecibo Message" toward a star cluster in the constellation of Hercules. When decoded, this first official radio message from earthlings to extraterrestrials creates a diagram showing the structure of DNA, the size of a human being, the planets of our solar system, and the Arecibo telescope itself.

Arecibo is also used as a powerful radar that electronically analyzes asteroids, determining which ones contain minerals and fuel sources that could be mined by a spacecraft. With a supply of fuel and materials available outside of the earth's deep "gravity well," humanity would be free to move outward into a self-supporting solar-system economy, spreading life permanently into space.

Max Planck Institute for Radio Astronomy Effelsberg, Federal Republic of Germany

The largest fully-steerable radio telescope in the world—100 meters across (far right)—teamed up with radio dishes in Great Britain and Australia to make this picture (upper left) of the entire celestial sphere as visualized from earth, in radio wavelengths. The bright band running the length of the picture is the Milky Way Galaxy.

This portrait of the radio universe demonstrates what astronomers are up against in the SETI effort: The radio sky is naturally full of complex emissions. Even so, contact between life on earth and life elsewhere in the universe is most likely to occur through giant radio dishes like the Effelsberg telescope.

The photo in the lower left shows life on earth as detected by satellites looking for chlorophyll in the ocean and for vegetation on land. When life is discovered elsewhere in space, we may gain perspective on our earthly ecological problems. Such an event would also accelerate the development of a global consciousness that will produce lasting solutions. Whatever the outcome, it is certain that we will never again view our species, nor the differences between races and cultures, in quite the same way.

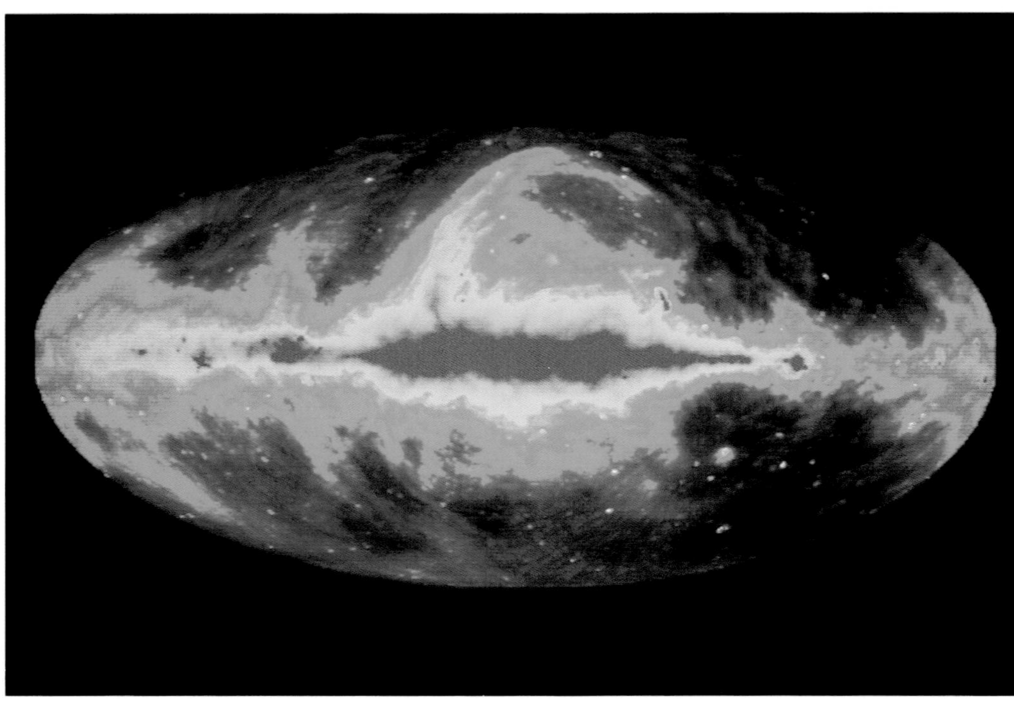

Photo © Max Planck Institute for Radio Astronomy

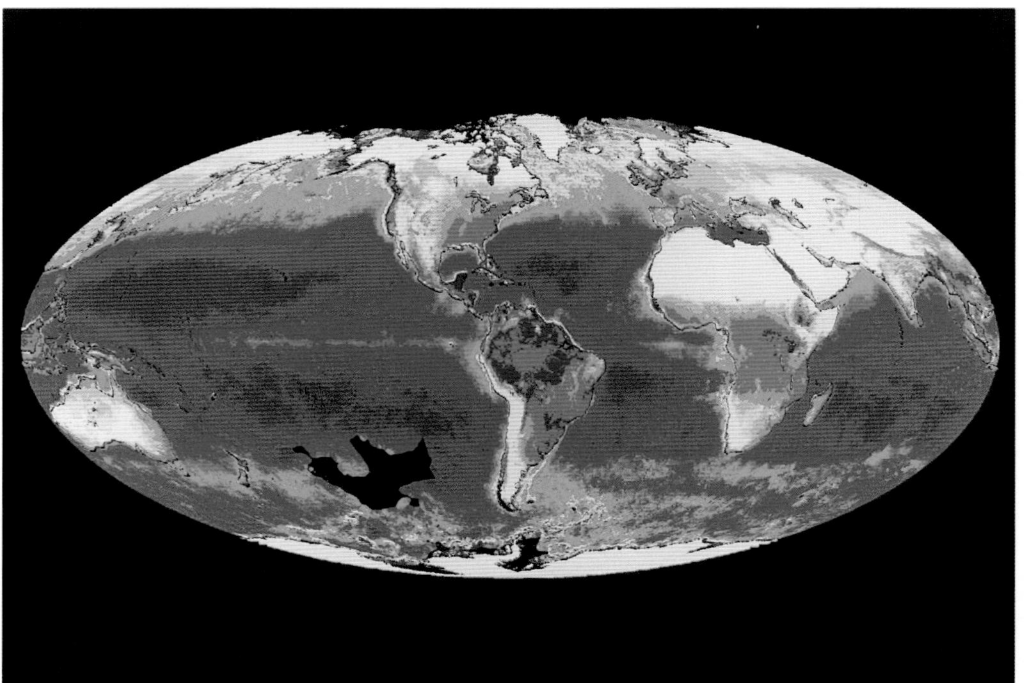

Photo courtesy Gene Feldman & Compton Tucker, NASA

I believe that this nation should commit itself to achieving the goal, before this decade is out, of landing a man on the moon and returning him safely to Earth. No single space project in this period will be more exciting, or more impressive to mankind, or more important for the long-range exploration of space; and none will be so difficult or expensive to accomplish. —John F. Kennedy, May 25, 1961

INTERNATIONAL ADVANCES IN SPACE EXPLORATION

Humanity stepped haltingly into the space age, propelled by the competition of the Cold War. It was the late 1950s, and satellite launches were motivated by the objective of proving, in the most public way possible, that the superpowers could reliably deliver heavy objects into earth orbit. The subliminal message conveyed by a successful flight was crushing: The same rocket that could fly into space could be directed against any nation on earth, carrying a cargo of atomic bombs.

The Soviet Union won all the early prizes in the space race: first and heaviest satellites in orbit, first lunar probe, first animal and first human in orbit. But in 1961 the momentum began to shift toward the United States. John Kennedy proclaimed the goal of landing a man on the moon by the end of 1969, and formally launched one of the most focused and inspired enterprises in history.

Kennedy's remarkable vision transcended military and political objectives. In the moon program he perceived a way to enrich the spirit of the entire nation. True, it would elevate the military capabilities of the aerospace community. But it would also stimulate America's educational process and inspire its young people to prepare for an adventurous future in space exploration. It would test the very limits of human insight and ingenuity with tasks of staggering complexity.

Kennedy's untimely death just 30 months later would only strengthen the nation's resolve. NASA's staff worked overtime, often without pay, to keep JFK's dream alive. It was a risky and daring gambit. When he first proposed the moon flight no American had ever been in orbit. The nature of the lunar surface was still unknown; many scientists believed it was covered with dust that would swallow any lander. To rendezvous and dock in orbit seemed hopelessly difficult even near earth—and a lunar mission would require doing so a quarter of a million miles away, while circling a lifeless orb without tracking stations.

There were other difficulties involved in this mission. America's biggest rockets were exploding on the launch pad, and the moon mission would require monstrous boosters that hadn't even been designed; new and unproven fuel-cell technology would be necessary to generate electricity, since batteries weren't sufficient for the eight-day trip to the moon and back; and many scientists believed the human body couldn't possibly survive long periods of zero-gravity.

Despite the seemingly impossible task it had set for itself, the United States pressed on into the future. In the process, the lessons of inspired leadership were learned again: give human beings a bold and interesting challenge that includes exploration into the unknown, and they will accomplish unforeseen and magnificent things.

Space exploration has benefitted humanity in ways that once only science fiction writers and futurists could imagine. It spurred the creation of small, powerful computers. It created satellite networks that have vastly improved the volume and quality of global communications. It lofted observational satellites that verify arms agreements, warn of devastating storms, and allow study of the earth's history, environment, and resources from a global perspective. Astronauts' pictures of earth, taken while en route to the moon, even helped awaken the ecological movement.

Apollo, as the moon program was named, succeeded in all of its objectives and inspired outstanding personal performance among the scientists, engineers and astronauts on the team. Today numerous nations are actively involved in space programs, and others are planning them. In the following pages we look at U.S. and other space programs worldwide, the technological torchbearers of the future.

NASA John F. Kennedy Space Center Florida, U.S.A.

Apollo 8 lifted off for man's first flight to the moon in December 1968, seven months before the first lunar landing. This flight was one of the most dangerous ever attempted. The Apollo spacecraft had flown only one manned mission, and was now being sent so far away that an emergency landing would take three days to accomplish. Upon its return home, the final countdown to the first lunar landing was on.

Photo courtesy of NASA

Gemini 6 photo by Thomas Stafford – NASA

Norton Air Force Base
San Bernardino, California, U.S.A.

The two-person Gemini spacecraft was flown ten times in the mid-1960s to develop techniques required for manned lunar missions. In his book Liftoff, *Apollo 11 astronaut Michael Collins aptly described Gemini as the bridge between Project Mercury and the moon.*

Gemini used a modified Titan II Intercontinental Ballistic Missile (ICBM) to achieve earth orbit. Twenty-five years later, the dependable Titan is still used for launching heavy satellites and interplanetary probes. Over 300 Titans have been launched to date.

In 1987, the last Titan II intercontinental missile was retired from active duty by the Air Force (far left). One by one, the Martin Marietta Corporation is refurbishing the old ICBM's for use as peaceful satellite launchers.

Gemini 7 photographed by Gemini 6
Low Earth Orbit

In a single joint mission flown in December, 1965, NASA overcame two of the biggest obstacles on its path to the moon. During Gemini 7's 14-day earth orbit, which proved human beings could survive weightlessness for the duration of a moon flight, Gemini 6 was launched in pursuit and successfully achieved the first orbital rendezvous between independently launched spacecraft.

Apollo 12 photo by Richard Gordon – NASA

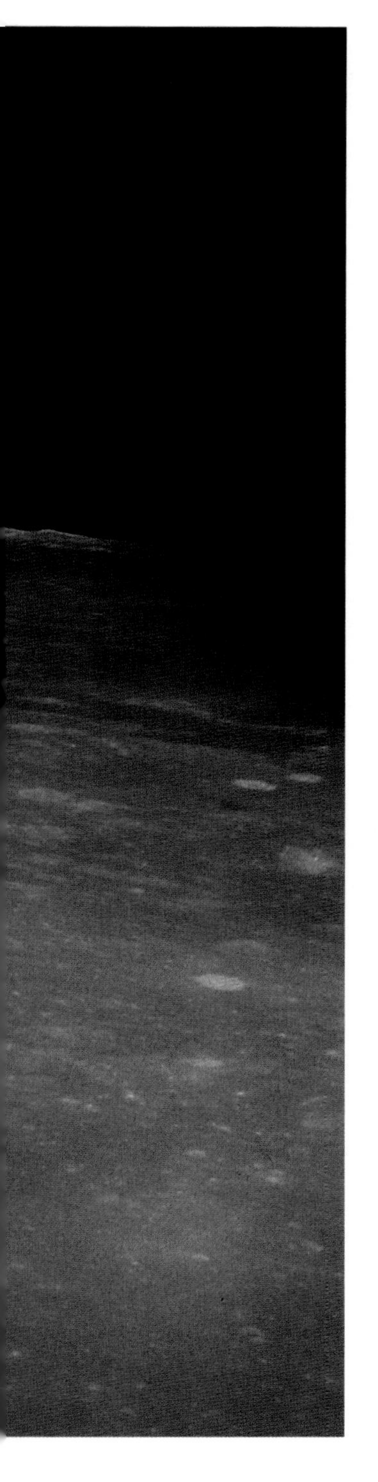

Apollo 17 photo by Harrison Schmitt – NASA

Lunar Module photographed from Command Module Lunar Orbit over crater Ptolemaeus

Lunar Module Intrepid, with Charles Conrad and Alan Bean aboard, prepares for history's second manned lunar touchdown. The moon's gravity, only one-sixth that of the earth's, enables this ungainly 15,000-kilogram spacecraft to make a rocket-powered round trip from lunar orbit to the moon's surface and back. On earth so much fuel is required to escape the force of gravity that it takes a vehicle weighing 10 times as much, the Titan II, to lift two astronauts on a one-way trip into orbit. Hence, as we continue our exploration of the solar system, it may ultimately become more economical for missions to use resources lofted from the moon than from Mother Earth.

Lunar Rover Sea of Serenity, the Moon

The lunar landings became more and more productive after Apollos 11 and 12. Astronauts had more time on the surface, and brought along wheeled vehicles to extend the range of their explorations. In a photograph taken by Harrison Schmitt, the first and only geologist to land on another world, mission commander Eugene Cernan prepares for a ride in a lunar rover. This December 1972 mission was Project Apollo's final trip to the moon. During four years of cislunar flight, 12 Americans had set foot on its surface; another 15 had flown by it.

In 1989, President Bush announced his commitment to a renaissance of human space exploration. His initiative seeks cooperation with Europe, Japan and the Soviet Union for visits to the moon and Mars.

McDonald Observatory
Mt. Locke, Texas, U.S.A. ●

The moon rocks collected by the Apollo missions will be studied for years to come, as will the data collected from instruments left behind on the lunar surface. One such device, a laser reflector, uses cube-shaped mirrors to return laser beams back to their source.

In this photo taken in west Texas, astronomers send blinding laser light to the moon. The returning beam will be used to measure the moon's orbit with previously impossible precision. Geologists use this same laser data, gathered simultaneously from both sides of an earthquake fault, to measure within inches the rate of continental drift on the earth's surface.

Lunar Eclipse
California, U.S.A. ●

The moon's orbital motion around the earth can be seen in this sequence of four pictures exposed on the same piece of film in August, 1989. The image was made over the course of 75 minutes, as the moon retreated from the earth's shadow during an eclipse. To make this photo, a camera with a telephoto lens was mounted "piggyback" on a telescope. The instrument's clock drive tracked the background of stars while the moon drifted through the foreground.

OVERSIZE

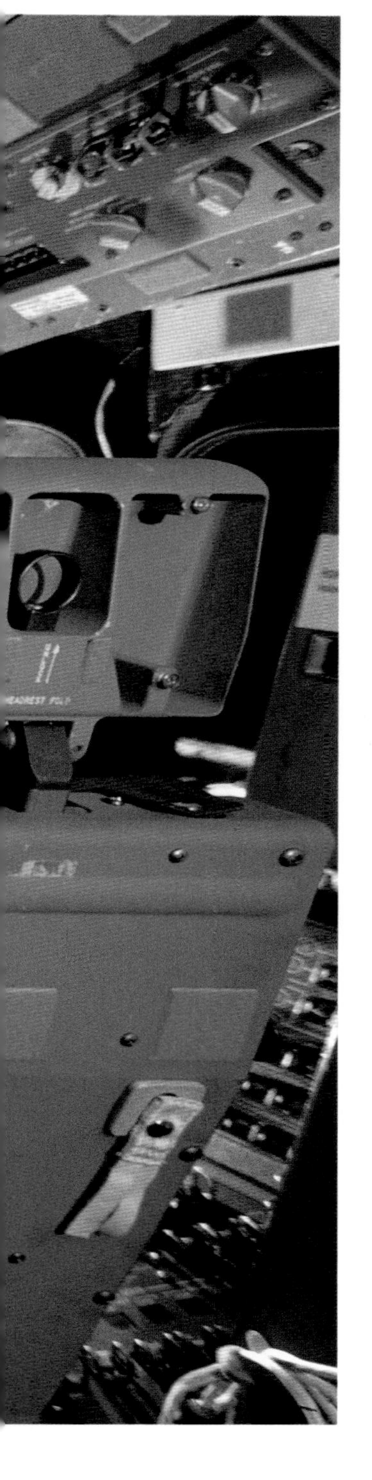

Space Shuttle Challenger
Lancaster, California, U.S.A. ●

(Previous pages) Challenger rolls into downtown Lancaster on the day it is delivered to NASA. The 1982 trip from the Rockwell assembly plant in Palmdale to Edwards Air Force Base required the upending of street signs and stop lights to make room for the orbiter's wings.

NASA Johnson Space Center
Houston, Texas, U.S.A. ●

(Left) Flight simulators in Houston enable shuttle pilots to hone their skills prior to actual space missions. Besides the hydraulic simulator used to create the sensations of banking and landing the winged orbiter, there are several stationary cockpit trainers like this one. Here astronaut Charlie Bolden sits in the mission commander's seat on the left side of the cockpit, while Steve Nesbitt rides in the pilot's chair.

Edwards Air Force Base
Edwards, California, U.S.A. ●

(Above) Quite a few graduates of the Air Force test pilot school at Edwards have gone on to become astronauts. T-38A jet trainers like the one being cleaned here by Airman First Class Mark Nowotny are used by NASA astronauts to keep their flying skills in shape. During training for specific shuttle missions, the jets enable the astronauts to commute efficiently between NASA centers and various contractor training facilities.

2
5

NASA Ames Research Center
Moffett Field, California, U.S.A. ●

Shortly after being selected as 1987's astronaut candidates, this group was sent on a whirlwind tour of NASA's facilities. Here they visit the gigantic wind tunnel at NASA Ames near San Jose. The trainees will be called "candidates" until they make their first flights in space and officially receive their astronaut wings.

NASA Johnson Space Center
Houston, Texas, U.S.A. ●

After full days of mission simulations, these spacemen spend their evenings playing rock and roll in an all-astronaut band called Max-Q. The name is NASA-speak for the "maximum aerodynamic pressure" that buffets space vehicles shortly after liftoff as they pass through the sound barrier. Shown here on July 20, 1989, at a 20th-anniversary party celebrating the first landing on the moon, are Max-Q astronauts James Wetherbee, Brewster Shaw Jr., Steven Hawley, Robert "Hoot" Gibson, and Pierre Thuot (left to right). The first live concert from space has yet to be booked.

NASA John F. Kennedy Space Center Florida, U.S.A.

When the crew of the 26th shuttle mission headed out to the launch pad in 1988, they were about to embark on a flight that would purge America's very soul. This was the nation's first manned mission since the Challenger explosion, 32 months earlier, claimed the lives of Christa McAuliffe, Dick Scobee, Mike Smith, Ron McNair, Ellison Onizuka, Gregory Jarvis, and Judy Resnik. Other astronauts who have given their lives include Apollo 1's crew of Gus Grissom, Edward White, and Roger Chaffee, who died in a launch pad fire in January 1967.

United States

Photo by Shuttle Mission 41-C – NASA

Photo by Shuttle Mission STS-5 – NASA

Space Shuttle Challenger
Low Earth Orbit

At far left, astronaut James "Ox" van Hoften floats through the empty cargo bay of the Challenger while on a satellite repair mission. The spacecraft has just made a successful rendezvous with the ailing Solar Max observatory. By replacing a few broken parts, van Hoften and fellow spacewalker George "Pinky" Nelson will extend the satellite's life by five years. Solar Max is the gold-plated object in the background.

Space Shuttle Columbia
Low Earth Orbit

Working in a comfortable shirt-sleeve environment on the shuttle's flight deck, astronaut Joe Allen (near left) prepares to take a photograph from the roof window. The window behind his head has a view into the cargo bay and is used to make pictures like the one at far left. Floating weightlessly alongside Allen's left elbow is a light meter.

NASA
923
Columbia
NASA
United States

Edwards Air Force Base
Edwards, California, U.S.A.

Two T-38A chase planes follow Columbia toward its triumphant first landing on the shuttle fleet's maiden voyage in April 1981. Few flights have been as gutsy as this one by John Young and Robert Crippen. Never before had anyone piloted the first flight of a totally new and untested rocket. Never before had anyone flown a winged vehicle on its maiden flight at hypersonic speeds, and with no engines. Despite a scare resulting from the loss of some protective tiles that had shaken loose during launch, the shuttle glided to a perfect landing on the dry lake bed at Edwards.

Risky flights like this one will always be a part of space exploration. It is astonishing that there hadn't been any in-flight deaths of American astronauts prior to the Challenger accident. Nevertheless, there have been many close calls. Gus Grissom nearly drowned in 1961 when his Mercury capsule sank into the ocean; John Glenn's Friendship 7 had an aerodynamically risky reentry due to a faulty heat shield indicator; Gemini 6 required a launchpad abort after *ignition of the Titan rocket, and Walter Schirra came dangerously close to ejecting himself and Tom Stafford from the cockpit; Gemini 8 spun wildly out of control while in orbit, and Neil Armstrong wrestled the spacecraft back into trim just moments before he became too dizzy to see straight. Armstrong had another close call on a later mission with Buzz Aldrin when Apollo 11's lunar module almost ran out of fuel during final approach for the first moon landing. And Apollo 13 nearly ended in disaster when the spacecraft's oxygen tanks ruptured en route to the moon, depriving the command module of its supply of breathable air and electricity. Jim Lovell, Jack Swigert, and Fred Haise barely survived, using the lunar module's oxygen supply and a remarkable series of makeshift emergency procedures.*

One thing is certain—men and women will die while exploring space. But the lessons of Apollo 1 and Challenger are clear: Even great tragedies cannot forestall our urgent yearning to explore the universe around us.

МИР КВАНТ ПРОГРЕСС М-1 СОЮЗ ТМ-8
ВИТОК 20788/14414/629/414/14
СУТКИ ПОЛЕТА
ЭКИПАЖА 27
ОБЪЕКТОВ 1321/917/41/27
НАЧАЛО ЗОНЫ 12:13:25
КОНЕЦ ЗОНЫ УСК 12:34:03
ДО НАЧАЛА ЗОНЫ 01:13:06
НАЧАЛО ЗОНЫ ЕВТ 13:45:11
КОНЕЦ ЗОНЫ УСК 14:10:49
СХЕМА ОБМЕНА ИНФОРМАЦИЕЙ

Kaliningrad Flight Control Center
Moscow, U.S.S.R. •

(Previous pages) In October 1989, flight controllers communicate with cosmonauts in the Mir space station as it passes over tracking stations on the Soviet mainland.

Following the triumph of Apollo, the U.S. space program sputtered along in the late 1970s and 1980s with budgets a fraction of what they were during the Apollo era. Meanwhile, the Soviet Union committed a far larger share of its national budget to space exploration.

Cosmonaut Training Center
Star City, U.S.S.R. •

(Above) Cosmonauts Anatoli Solovjev and Munir Habib, seated in a Soyuz spacecraft simulator, rehearse re-entry procedures for their return to earth. Habib, from Syria, is a participant in the Soviet Intercosmos program, which sends foreign space explorers into orbit for a visit to the Mir space station.

Baikonur Cosmodrome
Leninsk, U.S.S.R. •

(Right) June 6, 1985: Cosmonaut Vladimir Dzhanibekov lifts off on his fifth space flight, accompanied by Viktor Savinykh. This was perhaps the most dangerous mission ever attempted by the Soviet Union. The mission required docking their Soyuz spacecraft with the crewless Salyut 7 space station. Since February, Salyut had been tumbling in orbit, crippled and powerless. In a scene out of a science fiction movie, Dzhanibekov opened his mask in the dark and freezing space station to test the air.

He was later quoted in National Geographic *magazine: "When the air hit my face, I realized how bitterly cold the station was. Moisture from my exhalations froze in a tiny cloud around my face. Ice was everywhere—on the instruments, control panels, windows. Mold from past occupations was frozen on the walls.... Without the ventilators to circulate air, carbon dioxide from our exhalations hovered around us like a big ball. Our heads began to ache, our arms and legs grew sluggish.... Suddenly the lights turned on and ventilators started whirring. We realized the station was saved. We had worked nearly 24 hours—it was time to sleep."*

Cosmonaut Training Center
Star City, U.S.S.R.

Cosmonauts have inhabited the Mir ("Peace") space station almost continuously since it was launched in February, 1986, and have learned how to adapt to weightlessness for a full year at a time. As Mir expands with the attachment of new modules, it becomes, in part, an elaborate test platform for Soviet visits to Mars.

Here cosmonauts practice walking in space outside a full-scale mock-up of Mir's first module. Water-filled pools like this one near Moscow are called "neutral buoyancy simulators" because the cosmonaut (or astronaut) can hover effortlessly in any orientation once his inflated space suit is properly ballasted with lead. The slightest push can result in simultaneous spinning, yawing, and pitching motions, just as in the weightlessness of space.

The first orbital space walkers found it exhausting to hold steady in one position, and almost impossible to accomplish their tasks. The problem has been solved with elaborate systems of grips and foot restraints on the outsides of modern spacecraft.

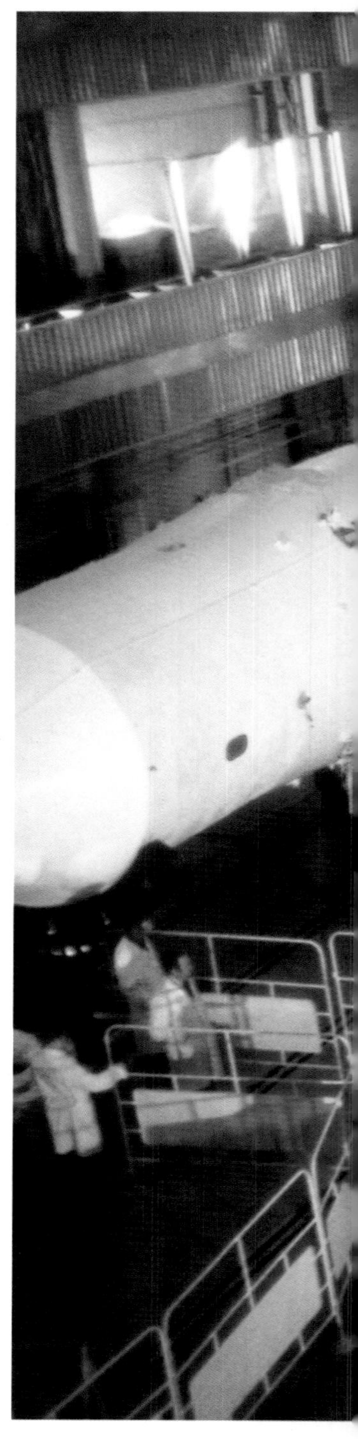

Cosmonaut Training Center
Star City, U.S.S.R. •

General Dzhanibekov visits his wife Lilia for lunch at their apartment inside the Star City complex. Today is his daughter Olga's 13th birthday. An accomplished artist during his off-hours, Dzhanibekov paints such subjects as sailors in space and human contact with extraterrestrials.

Baikonur Cosmodrome
Leninsk, U.S.S.R. •

Since Yuri Gagarin made man's first orbital flight in 1961, the Baikonur Cosmo-drome has been at the hub of the Soviet Union's manned space effort.

At right, space-suited cosmonauts Dzhanibekov and Gurragcha stroll through the SL-4 booster's assembly hall at Baikonur prior to the launch of Soyuz-39. The SL-4 launch booster is a direct descendant of the rockets that launched Sputnik, the world's first satellite, in 1957. It is still used to launch cosmonauts to Mir, from the same pad that Gagarin used.

One hundred rockets—five times the U.S. average—are launched into orbit every year from three Soviet cosmodromes, each far larger than the Kennedy Space Center. And while the U.S. discarded its mighty Saturn V along with the tools to build it, the Soviets created the giant Energia, by far the largest booster available today.

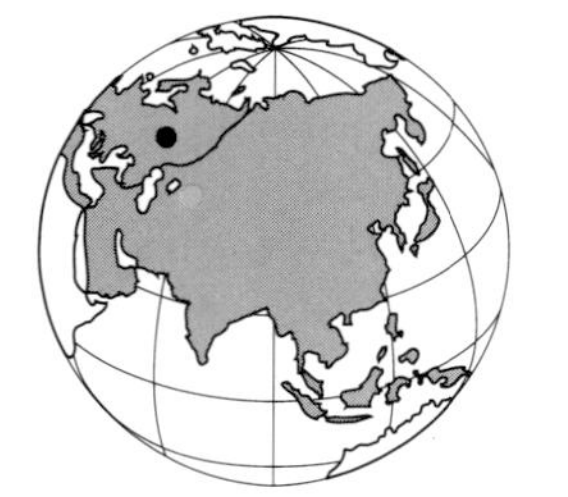

Photo by A. Mokletsov – © Novosti Press Agency

Xichang Satellite Launch Center
Xichang, China

China's family of Long March boosters has become a rugged competitor in the growing international market for commercial satellite launches. Due to the Long March's high success rate (96%) and a per-satellite cost that's one-third lower than in the West, the Chinese have been contracted by foreign customers to perform several launches in the early 1990s. Most will use Long March 3 and 2E rockets lofted from Xichang, one of four launch facilities in China.

The old and the new Chinas meet at Xichang's launch pad (left) in Sichuan Province. A mere 150 meters and one brick wall separate the launch pad from adjacent farm land. Farmers are required to stand back only one-half of a kilometer when rockets take off, despite the fact that the only Long March failure resulted in the rocket's crashing right next to the pad.

China's inland launch facilities haven't resulted in any human casualties, according to Chinese official Wang Yue at Shanghai's rocket factory. Says he, "We haven't had any trouble because the impact area is carefully selected. The only casualty was when a first stage killed a cow. The launch trajectories are approved by high officials, and we see no risk."

Despite such assurances, Western satellite manufacturers think it prudent to obtain guarantees from the Chinese government that they won't be held liable for any injury or property damage that occurs when their equipment is launched.

In the picture above, a launch crew rehearses at Xichang's command and control center. On the following pages, a Long March 3 undergoes testing.

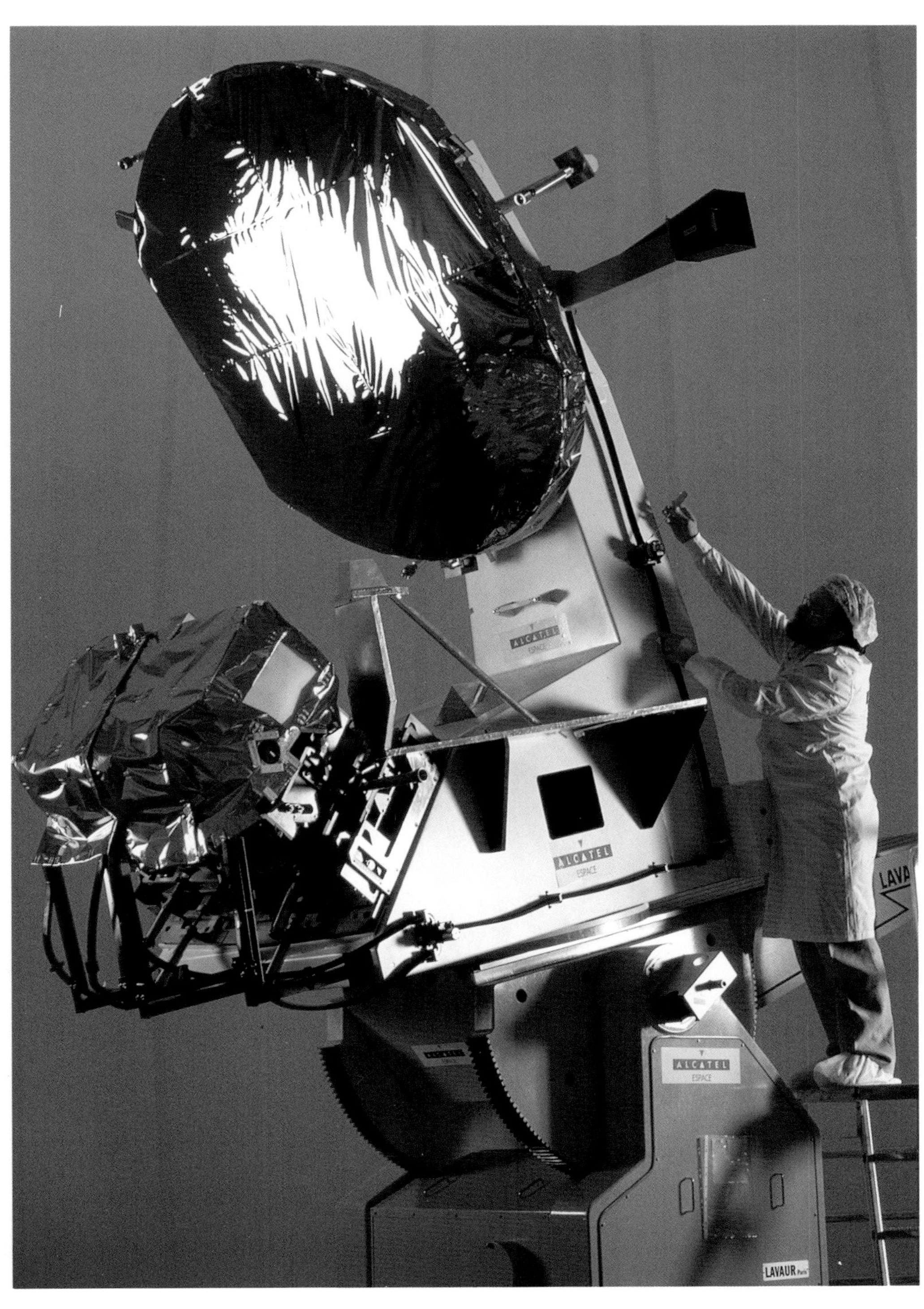

Alcatel Espace
Toulouse, France

Europe's commitment to space has expanded in recent years. Governments and companies are devoted to extensive multinational efforts that involve cooperation with both the Soviet and U.S. space programs.

The European Space Agency (ESA) was established in 1975 and now counts 13 European countries plus Canada as members. Europe's Ariane rocket performs more than half of the world's commercial launches.

The Toulouse plant of Alcatel Espace specializes in satellite circuitry and communications systems. The bulbous "radomes" atop the building (far left) are inflatable clean rooms that hold satellite antennas during pre-flight testing. Their plastic walls are designed not to interfere with radio transmissions between antennas and the distant calibration towers that measure signal coverage and strength. At near left, an antenna for the European communications satellite EUTELSAT II is prepared for radome testing.

aerospatiale
aerospatiale

Aerospatiale
Les Mureaux, France

Aerospatiale oversees a lion's share of the Ariane rocket program on behalf of CNES, the French space agency responsible for the effort. Parts from contractors across Europe converge for final assembly and testing at Aerospatiale's plant in Les Mureaux, near Paris. At far left, final wiring is completed on the first stage's propulsion bay above the rocket motors. At near left, engineer Jacky Janik tests third-stage engine electronics.

Aerospatiale's Shirley Compard explains why Ariane has earned so much launch business (over 10 satellites a year): "In the U.S. there are yearly appropriations squabbles in Congress. But when decisions are made by European governments, funding is guaranteed for five or ten years. We know how to cooperate in Europe, because if we don't cooperate, we can't accomplish anything. The continuity makes us less expensive in the long run." Recently, the European Community ordered dozens of new Ariane 4 launchers, a volume that guarantees continued low launch prices. And the Ariane 5, scheduled for use in the mid-1990s, is designed to launch larger satellites at an even lower cost per kilogram.

Société Européenne de Propulsion (SEP)
Vernon, France ●

SEP is the leading rocket engine company in Europe, and a partner in Ariane's success. Four hundred of its Viking first- and second-stage engines have been built, and 400 more are on order for Ariane 4 flights scheduled in the 1990s.

At the plant's front gate, a model of the engine is decorated annually for Christmas (near right). SEP also builds and tests the new Vulcain cryogenic engine, which uses superchilled liquid oxygen and liquid hydrogen as propellants. Designed for the first stage of the upcoming Ariane 5, these fuels are more efficient but harder to handle than earlier materials. The older Viking, like the U.S. Titan and the Chinese Long March rockets, uses room-temperature hypergolic fuels that ignite on contact with each other.

Guiana Space Centre
Kourou, French Guiana ●

At far right, firefighters at Ariane's launch pad in South America wear protective garments to guard against flames, fumes, corrosion, and explosions that could occur around a fully fueled rocket. Fortunately, the outfits have never been put to the test in an actual emergency.

Camiva
23
SAPEURS POMPIERS
cnes
DE PARIS

Guiana Space Centre
Kourou, French Guiana

An Ariane 4 leaps skyward during a time exposure, its flaming streak interrupted by low-level clouds. Stars trail in the background; the strap-on boosters are barely visible, tumbling back to earth after exhausting their supply of fuel. Launches here are normally scheduled for the evening, when the volatile tropical climate is most likely to be favorable for liftoff.

Technician Jean-Luc Bresson, at right, wears an airtight ERGOL suit while fueling Ariane's first stage. Fumes from the hypergolic fuels are extremely toxic, and launch spectators several kilometers away are given gas masks in case of an accident.

Both local geography and the laws of physics make French Guiana an ideal launch site. Only one pad is necessary for both easterly geosynchronous trajectories and north-south polar orbits. Thousands of kilometers of open ocean to the east and north ensure that falling rocket stages won't land in populated areas. In order to achieve the same polar and equatorial capabilities, the U.S. requires launch pads in both Florida and California.

The French Guiana site derives an added advantage from its proximity to the equator, where the rotational velocity of the earth's surface imparts an extra push of almost 1,700 kilometers per hour to rockets headed eastward. The added boost means a given rocket can launch 10% more payload from French Guiana than from Cape Canaveral. To achieve those same benefits, the U.S. is planning a new launch facility in Hawaii, the state closest to the equator, and the only one that offers unobstructed trajectories toward both polar and equatorial orbits.

SN SOCOCER

Tanegashima Space Center
Tanegashima, Japan ●

(Previous pages) New construction at Tanegashima, probably the most beautiful launch site in the world, proves Japan's commitment to long-range space adventure, exploration, and profit. Here, an H-1 satellite booster awaits a "go" for launch. Directly behind the rocket, a construction crane puts the final touches on a new gantry that will launch the H-2, an all-Japanese rocket with more than three times the capacity of the H-1. Scheduled for testing in 1992, the H-2 will free the Japanese from competitive restrictions in their license to adapt the H-1 from the U.S. Delta rocket. Head-to-head competition between the H-2 and other launch services worldwide is likely to result.

Ishikawajima-Harima Heavy Industries Co., Ltd. (IHI)
Tokyo, Japan ●

(Near right) These two racks of Japanese experiments, aimed at manufacturing valuable new materials in space, will fly on a U.S. shuttle mission in 1992.

Mitsubishi Heavy Industries, Ltd. (MHI)
Nagoya, Japan ●

(Far right) Two of the final four H-1 rockets on order undergo assembly and testing. Larger H-2 fuel tanks in the background await the imminent conversion of the plant into an H-2 facility.

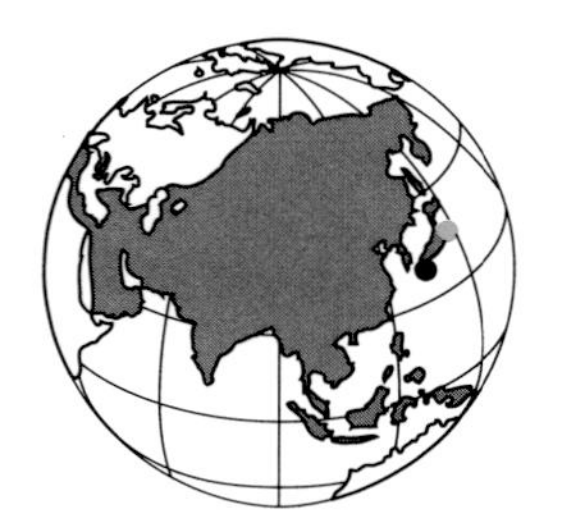

***NASDA Tsukuba Space Center Tsukuba, Japan* ●**

Technician Kenichi Saga monitors a launch from the mission tracking and control room at the NASDA space center. NASDA, the National Space Development Agency of Japan, coordinates launch efforts using H-1 and H-2, while a second Japanese space agency called ISAS uses its own series of rockets for scientific missions to the moon and beyond.

***Tanegashima Space Center Tanegashima, Japan* ●**

September, 1989: The H-1 performs NASDA's 20th successful satellite launch, lofting a new weather satellite to monitor the Pacific. Before this launch, Tanegashima Island was battered by typhoons that resulted in several launch delays.

Japan's robust space program is a model of efficiency. It has accomplished much on a relatively modest billion dollars a year, a budget less than one-tenth of NASA's.

In conjunction with America's space station project, Japanese contractors are currently building both free-floating and attachable modules capable of manufacturing sophisticated metals and crystals in space. An unmanned space plane will fly those products back to earth, independent of America's shuttle.

H20
NIPPON

Photo by Voyager 1 – NASA / JPL

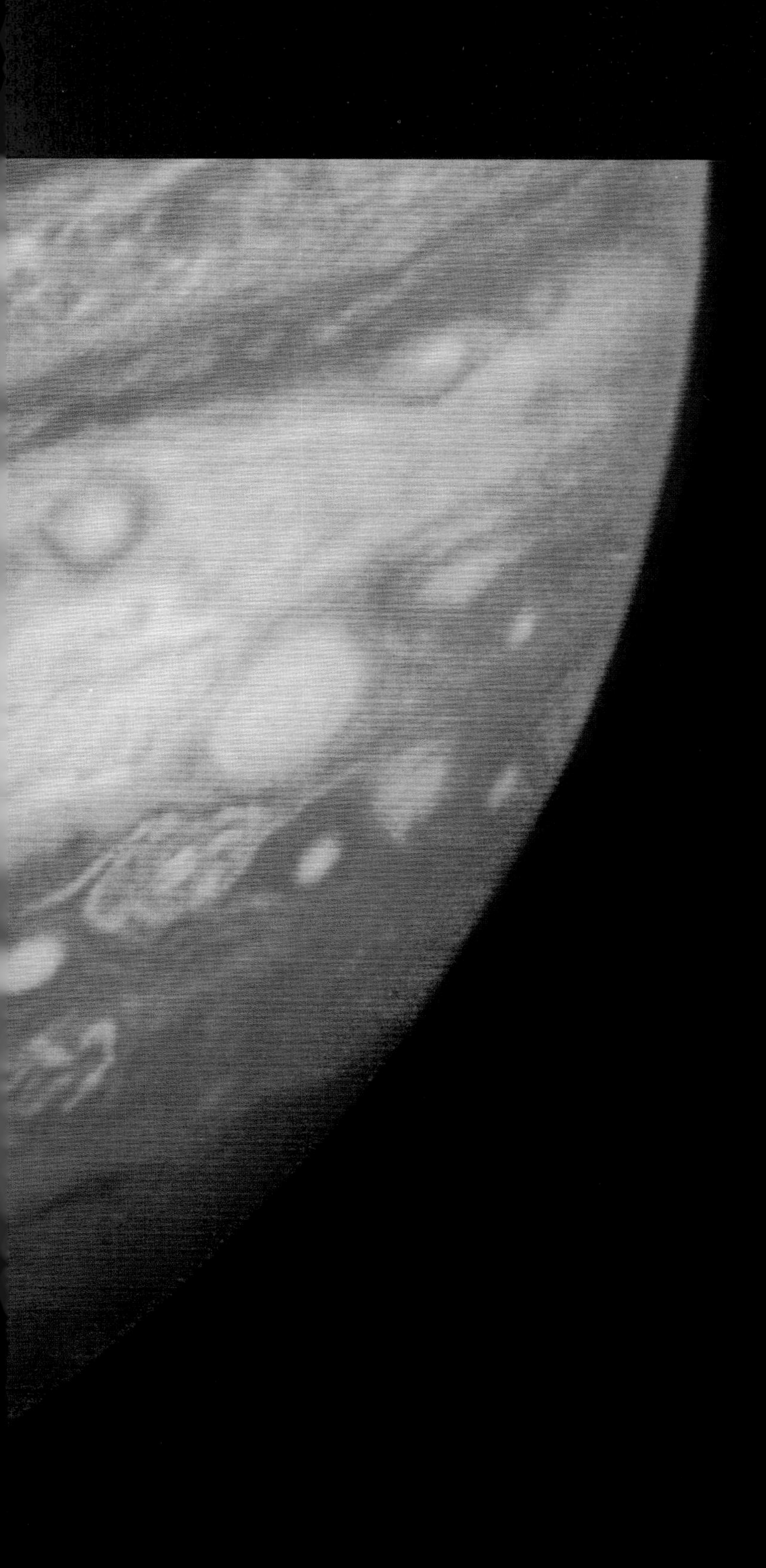

Today we can no more predict what use mankind may make of the Moon than could Columbus have imagined the future of the continent he had discovered. —Arthur C. Clarke

Since the beginning of the space age, robotic explorers have routinely preceded us into hostile new environments, serving as our eyes and ears, measuring and defining the risks of manned travel, and clearing the way so that astronauts and cosmonauts could safely follow.

Interplanetary exploration began close to home in 1959, when the Soviet Union's crewless Luna 3 spacecraft sent back the first fuzzy photos of the far side of the moon, which is never visible from earth. A few years later, America's Ranger and Surveyor robots determined, contrary to some scientists' beliefs, that the moon's surface wasn't covered with too much dust for us to land there. Then the Lunar Orbiters mapped the moon's surface in detail, enabling engineers and geologists to compromise on landing sites for Apollo that were both safe and scientifically interesting.

In a mere thirty years our robots have reached beyond the moon and visited most of the solar system, starting with the inner planets. American Mariners and Pioneers reconnoitered Mercury and Venus, and two Viking landers set down on Mars. Early versions of the Soviet Venera spacecraft flew by Venus, while later Veneras orbited and landed there. More nations joined in the exploration: The French-Soviet Vega went on from Venus to Halley's comet, and Japan has launched one lunar and two comet probes of its own.

The outer planets were the targets of four American spacecraft launched in the 1970s: Pioneers 10 and 11, and Voyagers 1 and 2. Each successfully accomplished its mission, and by 1989 robots had visited Jupiter, Saturn, Uranus, and Neptune, and were headed toward the farthest fringes of the solar wind and into interstellar space.

The Voyager 2 mission to four separate planets was the most remarkable of them all, for it demonstrated humanity's mastery of gravity-assisted trajectories. Voyager 2's economical Grand Tour of the outer solar system, made possible by a planetary alignment that occurs only once every 176 years, used the gravitational force of each planet as a celestial slingshot to hurl itself toward its next target. The technique enabled the probe to travel in directions and at speeds that were far beyond the capabilities of its rocket engines.

One of the most thrilling results of the Voyager missions has been our growing knowledge of the moons that surround the outer planets. In time we will know these satellites far more intimately than the planets themselves. Because of their light masses and low gravities, it will take little fuel to land on them and return to orbit. Also, these moons have hard surfaces and potentially useful mineral resources, whereas the outer planets themselves are gaseous.

Since Galileo pointed his telescope at Jupiter, over 60 new moons have been discovered in the solar system, 22 of them by the two Voyagers. The smallest moon known is tiny Deimos, an oblong rock only 15 kilometers long which orbits around Mars. At the other end of the scale, five satellites are larger than the earth's moon.

From 1978 to 1989 there were no interplanetary missions launched by the U.S. Today, however, we are entering another golden era of planetary exploration. Space probes have recently been sent to Venus and Jupiter, and others are being built. Moreover, the whole world is getting involved: New spacecraft normally carry experiments provided by several nations.

Planetary exploration started with earth. America's first satellite, Explorer 1, reported in 1958 that the planet is surrounded by radiation belts that astronauts would need to avoid, thereby proving the value of robotic reconnoitering. Today, a series of new probes called "Mission to Planet Earth" is scheduled for the mid-1990s. These advanced satellites will study the earth's environment and resources in new ways, so that we can assess the damage we've done during the 20th century, develop solutions for the earth's environmental problems, and prevent new problems from arising.

Skeptics say we should give up expensive and dangerous manned exploration, and rely instead on robots to explore the universe. But in so doing, we would lose the spirit of exploration that invigorates and motivates the entire human race. True, the unmanned planetary missions thrill scientists, but they don't inspire average citizens in the same way. More importantly, robots have neither intuition nor adaptability but can only seek what they are programmed to find. Nor can they convey the thrill of discovery, or the sense of wonder and awe. While most of us remember where we were when Neil Armstrong and Buzz Aldrin first walked on the moon, for example, few remember that shortly thereafter the Soviets sent robots to the moon to collect their own surface samples.

Once the mining of resources from the solar system becomes profitable, the cost of manned exploration will be irrelevant. Even then, as we begin sending robots to other solar systems, we will probably continue to debate whether it is worth the cost of sending humans as well. The answer, I suspect, will be the same as always: Onward!

Space Shuttle Atlantis
Low Earth Orbit

The Magellan spacecraft, with its antenna at top and interplanetary rocket engine at bottom, is launched from the cargo bay of Atlantis in May, 1989. This was the first planetary probe to be dispatched by the United States in 11 years. Magellan's rocket was fired shortly after this picture was made, propelling the craft out of earth orbit and toward a 1990 injection into orbit around Venus. While in orbit, Magellan will use a powerful radar to penetrate the planet's perpetual cloud cover and produce detailed maps of the surface.

Photo by Shuttle Mission 30 – NASA

USA

Jet Propulsion Laboratory (JPL)
Pasadena, California, U.S.A.

JPL is the nerve center for all of America's (and most of the world's) interplanetary probes. Since the early 1960s, JPL has handled tracking, navigation, control, and data acquisition for missions to the moon and beyond. It communicates with distant probes through a global system of dish antennas, known collectively as the Deep Space Network, or DSN. Three stations using 70-meter receivers are spaced evenly around the globe, in California, Spain, and Australia, and can therefore provide continuous 24-hour coverage of any point in the solar system. Even Soviet missions use JPL and the DSN for guidance.

JPL scientists design, test, and operate most of America's planetary missions. The Galileo spacecraft (left) was launched toward Jupiter by the space shuttle Discovery in October, 1989. The trip will take six years, during which Galileo flies past Venus once and Earth twice, using the gravitational force of those planets to accelerate the craft on its way. This complicated and lengthy flight plan became necessary when, in Challenger's aftermath, a more powerful—and dangerous—liquid-fueled upper stage was disqualified from flying in the space shuttle.

Galileo's mission is not a brief flyby like the ones that preceded it. It will be the first spacecraft to orbit Jupiter, keeping the planet and its moons under constant surveillance for two years and launching the first probe into Jupiter's atmosphere.

To test the craft's space-worthiness, NASA engineers "flew" it inside the giant vacuum chamber in the left photo, bombarding it with intense ultraviolet light to simulate the fierce temperatures it will encounter during its closest approach to the sun. The light is so intense that without this protective outfit, human skin would fry in seconds.

(Above) JPL's Space Flight Operations Center is the focal point of the DSN. It is currently managing both Magellan and Galileo, as well as the older Voyagers and Pioneers that are still transmitting data from the fringes of the solar system.

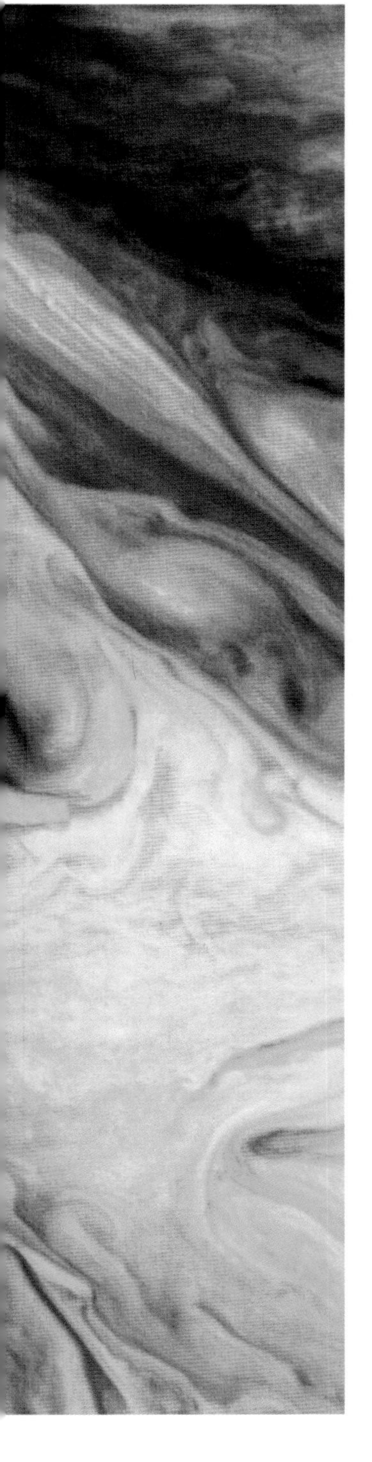

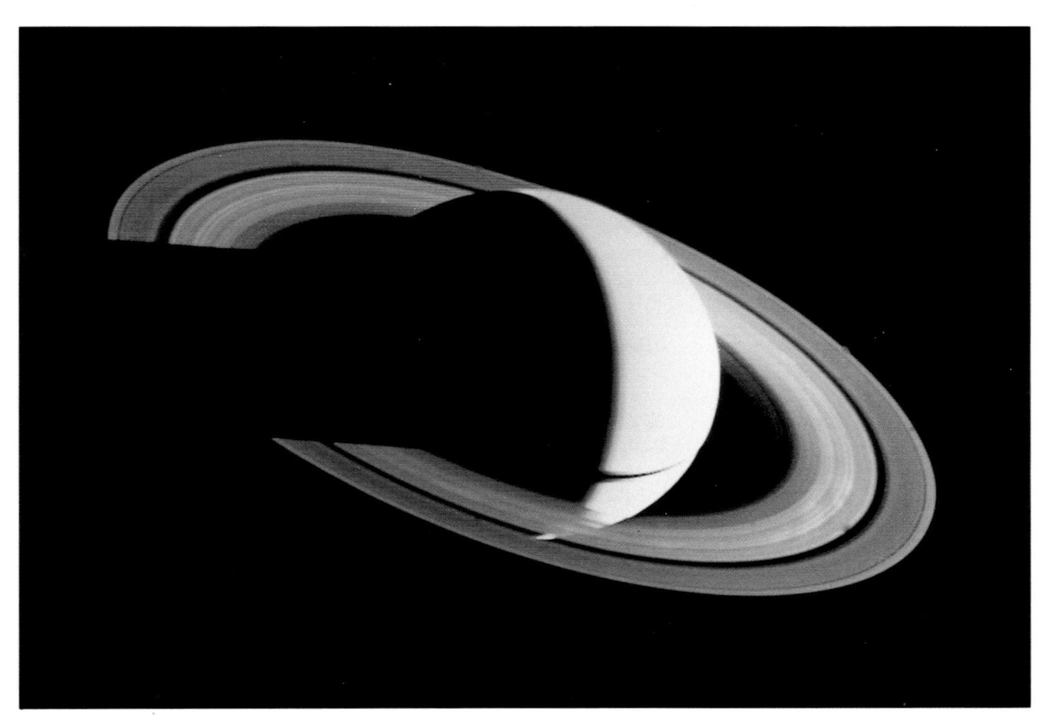

Photos by Voyager – NASA / JPL

Jupiter and Saturn photographed by Voyager

Combined, Voyagers 1 and 2 have flown past all of the outer planets except Pluto. Launched in 1977, each spacecraft encountered first Jupiter and then Saturn. At Saturn their paths diverged, with Voyager 1 leaving the plane of planetary orbits to fly by Saturn's largest moon while Voyager 2 continued on to Uranus and Neptune.

(Far left) Jupiter's diameter is 11 times greater than earth's, and its mass is more than twice that of all the other planets in the solar system combined. If it were a little larger, it would have become a star. The stormy, turbulent structure of Jupiter's Great Red Spot is evident in this false-color enhancement of an image made in 1979. The spot is so large that several objects the size of earth could be dropped into it.

Jupiter has four large, bright moons discovered by Galileo in 1610, and 12 smaller ones. Scientists were amazed when Voyager cameras revealed that one of the Galilean moons, Io (pronounced EE-oh), has a smooth surface with no craters at all. Photographs of Io, like the one at top left, showed actual volcanic eruptions in progress, suggesting that Io's craters have been filled with ejecta.

(Bottom left) Voyager 1 made this "farewell" photo of Saturn as it departed.

Jet Propulsion Laboratory (JPL)
Pasadena, California, U.S.A.

Voyager 2 passed by Saturn in 1981 and used Saturn's gravity to accelerate itself toward Uranus and Neptune. During its final flyby on August 24, 1989, Brad Smith (left), head of Voyager's imaging team, and Ed Stone, chief Voyager scientist, study Neptune's Great Dark Spot on a monitor at JPL's Image Processing Lab. Said Stone, "I don't see how any of us could be any happier. These data form nothing less than the encyclopedia of the planets."

At 4:00 A.M., while receiving photos of Neptune's large moon Triton, Stone (right, foreground) joined an impromptu discussion among scientists that included Larry Soderblom (left) and Carl Sagan (center). Despite the late hour, spirits were soaring. Said JPL's Jurrie van der Woude, "The Neptune encounter was one fabulous surprise after another, with Neptune being indescribably beautiful, and Triton being a geological stunner. It was extremely emotional, but with the joy there was also a sadness among the team members. In our lifetimes, this was probably the last planet that will be seen up close for the first time. How do you beat that?"

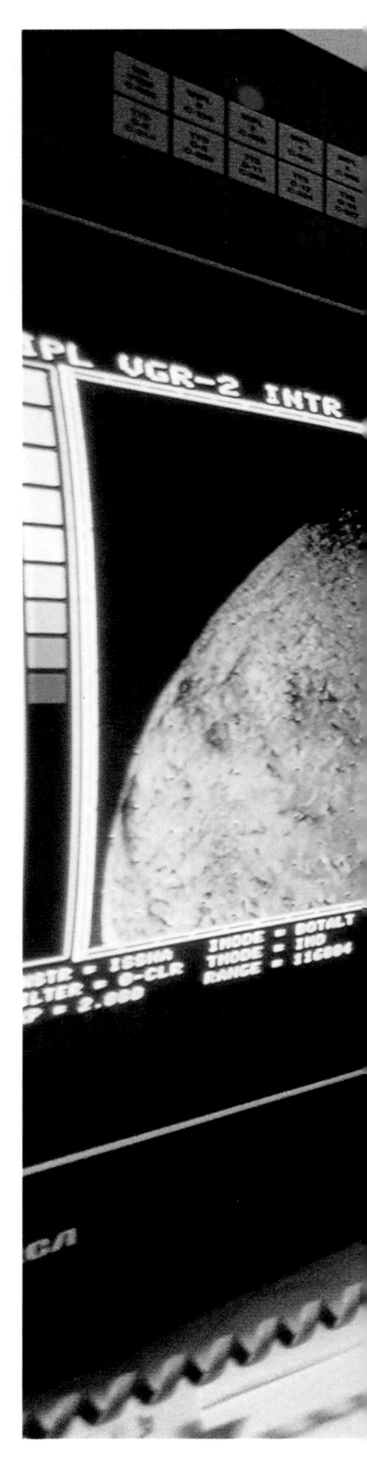

Photo by Voyager 2 – NASA / JPL / USGS

Voyager's Final Encounter
Triton, near Neptune

Triton, the coldest place yet visited by a spacecraft from earth, surprised scientists with its varied geography and active volcanoes (above). U.S. Geological Survey specialists combined dozens of photographs to make this detailed image, using a computer to remove blurring, distortion, and seams.

Jet Propulsion Laboratory (JPL)
Pasadena, California, U.S.A.●

That same day, as live pictures streamed in from Neptune's satellite Triton, Jeffrey Cuzzi used an interactive computer to enlarge and enhance interesting regions (right). The sun was rising outside, and Cuzzi hadn't slept for 24 hours, but the thrill of discovery kept him going for hours more.

The Very Large Array (VLA)
Socorro, New Mexico, U.S.A.●

(Following pages) So feeble were the signals from Voyager 2 when it reached distant Neptune that steps were taken on earth to improve the data link. The three Deep Space Network (DSN) antennas were enlarged from 64 to 70 meters, and linked up with nearby radio telescopes. Returning data was then fed into one central computer that combined, compared, and filtered the information.

The California DSN antenna was linked with the 27-dish VLA in New Mexico, combining the signals from Voyager into a "much louder whisper," as one scientist put it. In Australia, the DSN antenna at Tidbinbilla was arrayed with the Parkes Radio Telescope. Using these combinations, scientists are hopeful that Voyager's whispers will be audible well into the next century.

237:10:44

NASA

If anyone thinks that this idea is fantastic ... I remind him that half a century ago a single man in the Atlantic sky was headline
How many thousands are up there at this very moment, dozing through the in-flight movie? —Arthur C. Clarke

T H E   D A W N   O F   T H E   S E C O N D   S P A C E   A G E

There is no question that we are at the beginning of a second era in space travel, one that promises ever greater benefits to our home planet. With growing momentum, the United States and other spacefaring nations are preparing for a permanent manned presence beyond earth. President George Bush's commitment to renewed lunar landings and missions to Mars, expressed on the 20th anniversary of the first moon landing, did not set deadlines as did John F. Kennedy's moon pronouncement of 1961. But his proposal does call for a program with permanence, putting an end to the relatively short-term approach that characterized the Apollo program.

For years, Apollo 11's Michael Collins has lobbied for Mars to be our next objective—a goal long embraced by the Soviet Union, and now being accepted by U.S. space planners as well. Collins' conviction is that no goal short of Mars will bring unifying purpose to NASA's disparate efforts. Certainly no other goal can offer the simultaneous benefits of research into efficient energy consumption, advanced recycling techniques, more durable products, new propulsion systems, adaptation to prolonged weightlessness, artificial gravity, bioregenerative life support systems, miniaturized medical facilities, and permanent space stations.

But why not forget about Mars, and focus on the moon instead? It's close, and it's a promising source of new metals, oxygen, and fuel. Or would it be more efficient to mine asteroids, or even the tiny moons of Mars? The orbital trajectories of these missions might require less energy than missions to and from the surface of our massive moon.

During the second space age we will cross a threshold beyond which space travel becomes an economic necessity, just as once occurred with air travel. But that change will only come about when enough equipment and people are permanently off-earth that the space economy can begin to sustain itself. Fuel will be gathered from afar, not lofted at great expense from earth. Equipment will be built with materials collected throughout the solar system. And little by little, our permanent presence in space will be assured.

A solar-system economy will have to be international in nature; start-up costs of any permanent program will be high, and joint efforts will be required to avoid costly duplication of technologies. But the benefits are sure to extend beyond economics, leading to greater international harmony as well. My hope is that the first international mission to Mars will act as an example to children everywhere of how to work together as one planet and one species. Inspired by the high drama of interplanetary internationalism, those children will soon grow up and make miracles happen on earth as well.

But first, decisions must be made. To the moon, or to Mars? The question is irrelevant in the long run, because we are destined to go both places. For the short term, though, either goal will require solving problems that relate directly to the survival of planet earth—our home spacecraft. In this final chapter, we take a look at new technologies for manned missions to the moon and planets.

Earth photographed from Apollo 17 Command Module Lunar Orbit

This picture of a crescent earth rising over the lunar crater called Ritz was taken in 1972, during Project Apollo's final mission to the moon. Before long we will visit the moon again, with a greater sense of purpose than ever before.

Apollo 17 photo by Ronald Evans – NASA

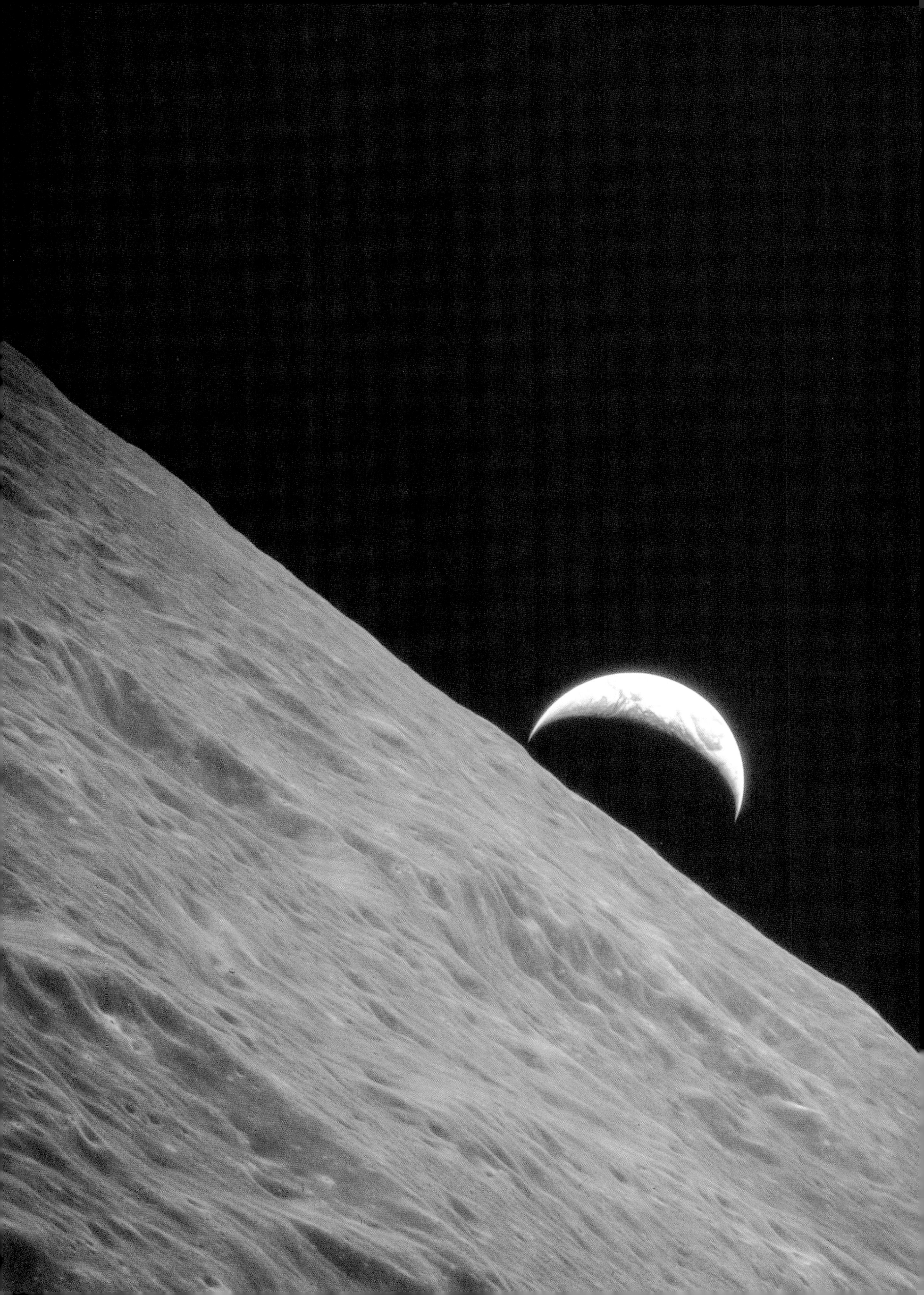

Photo by François Colas, © Pic du Midi

Pic du Midi Observatory
Pic du Midi de Bigorre, France

It is important to keep studying Mars if we are going to visit there. NASA will launch a probe called the "Mars Observer" in 1992; the Soviets are planning to send drilling robots, balloons, orbiters, and rovers in 1994; and a Soviet sample-return mission, a dress rehearsal for an actual manned landing, could take place as early as the year 2000.

From earth, it has always been very difficult to make good photographs of the planets because our turbulent atmosphere distorts the view—witness the way stars "twinkle" at night. Still, earthbound observations provide essential information. Pic du Midi Observatory, at an altitude of 2877 meters in the French Pyrenees, gets a lot of bad weather, but for a few nights every year the seeing is remarkably steady. During such moments, telescopes here have made some of the sharpest earth-based images of the moon and planets. An example of their very best work is this electronic CCD image of Mars (above), made in 1988. The white spot at top is the planet's ice cap.

From space places like this we have learned that Mars has a day about 24 hours long and four seasons just like earth. Its atmosphere is thin and offers little oxygen. Temperatures are earth-like during the day, but intensely cold at night.

NASA

NASA Langley Research Center
Hampton, Virginia, U.S.A.

Fuel-saving "aeroshells" will do some of the work of rocket engines on interplanetary flights of the future. "Aerobraking" involves grazing a planet's atmosphere to let it absorb the excess speed from a planetary approach. Air friction saves tons of fuel and enables larger payloads to be sent on interplanetary missions. When aerobraking is complete, the craft, "captured" by the planet, will skip back out of the atmosphere into orbit. The aeroshell itself is a giant heat shield which protects the spacecraft during the plunge.

The concept will be tested soon by a satellite called the Aeroassist Flight Experiment, or AFE. Once in orbit, AFE's powerful rocket engine will fire it into earth's atmosphere at the approximate speed of an approach from interplanetary space. Its oblate aeroshell will enable mission controllers to "fly" it along the upper atmosphere. At left, NASA's Tom Campbell positions a model of the AFE in an echo-free chamber that simulates the radio conditions of space. His data will determine the best placement of the satellite's antennas.

At NASA Langley's Spacecraft Analysis Branch, youthful computer wizards like Gary Qualls (above) conceive futuristic space stations and planetary vehicles. Qualls's advanced space station, shown here, uses a rotating ring for artificial gravity, and a stationary cargo bay for assembling a Mars vehicle with two aeroshells.

NASA Marshall Space Flight Center Huntsville, Alabama, U.S.A.

The final assembly and testing of interplanetary vehicles will take place at orbiting spaceports. The ports will require large trusses to hold together a collection of habitation modules, laboratories, tank farms for fuel and supplies, and solar panels. Building the trusses will be easy thanks to experiments conducted under the guidance of Walter "Doug" Heard at NASA Langley. Above, in NASA's largest underwater weightlessness simulator, space-suited technicians practice truss fabrication using spars with interlocking tips. The workers ride on moving platforms called Mobile Transporters—the space-age equivalent of hydraulic cherry pickers.

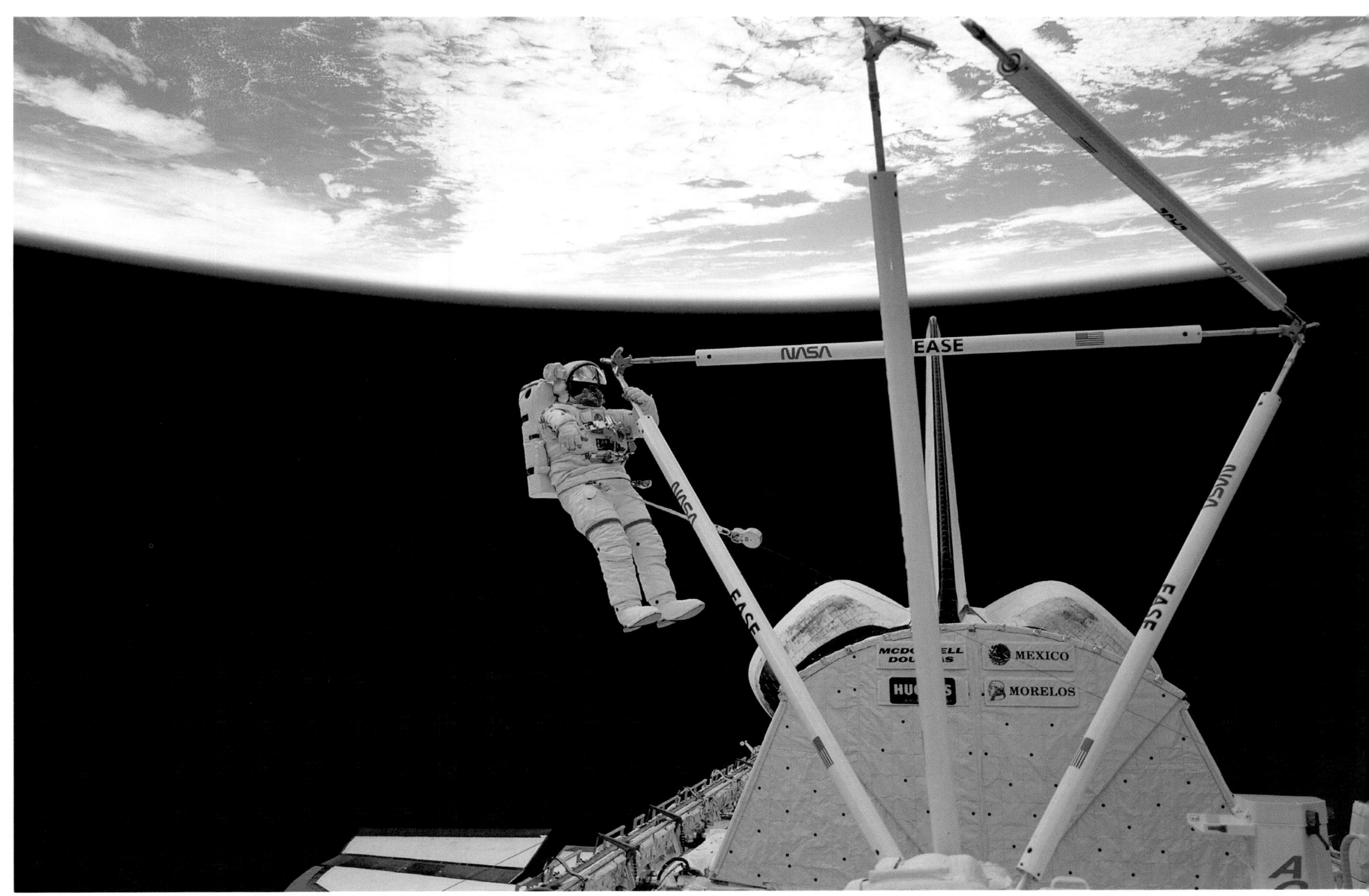

Photo by Shuttle Mission 61B – NASA

Space Shuttle Atlantis
Low Earth Orbit

Doug Heard's prototype space station trusses have already been tested in orbit. In late 1985, astronaut Jerry Ross floated above the shuttle's cargo area and joined interlocking beams into this rigid, triangular lattice.

Starting in the mid-1990s, the United States will begin the step-by-step assembly of space station Freedom. When finished, the multi-national spaceport—its modules owned by the U.S.A., Europe (with Canada), and Japan—will facilitate zero-gravity manufacturing as well as biomedical research into the effects of extended weightlessness.

Eventually, Freedom or its off-spring will include vehicle-assembly hangars. A system of space tugs will ferry satellites and supplies to and from higher orbits in the earth/moon system.

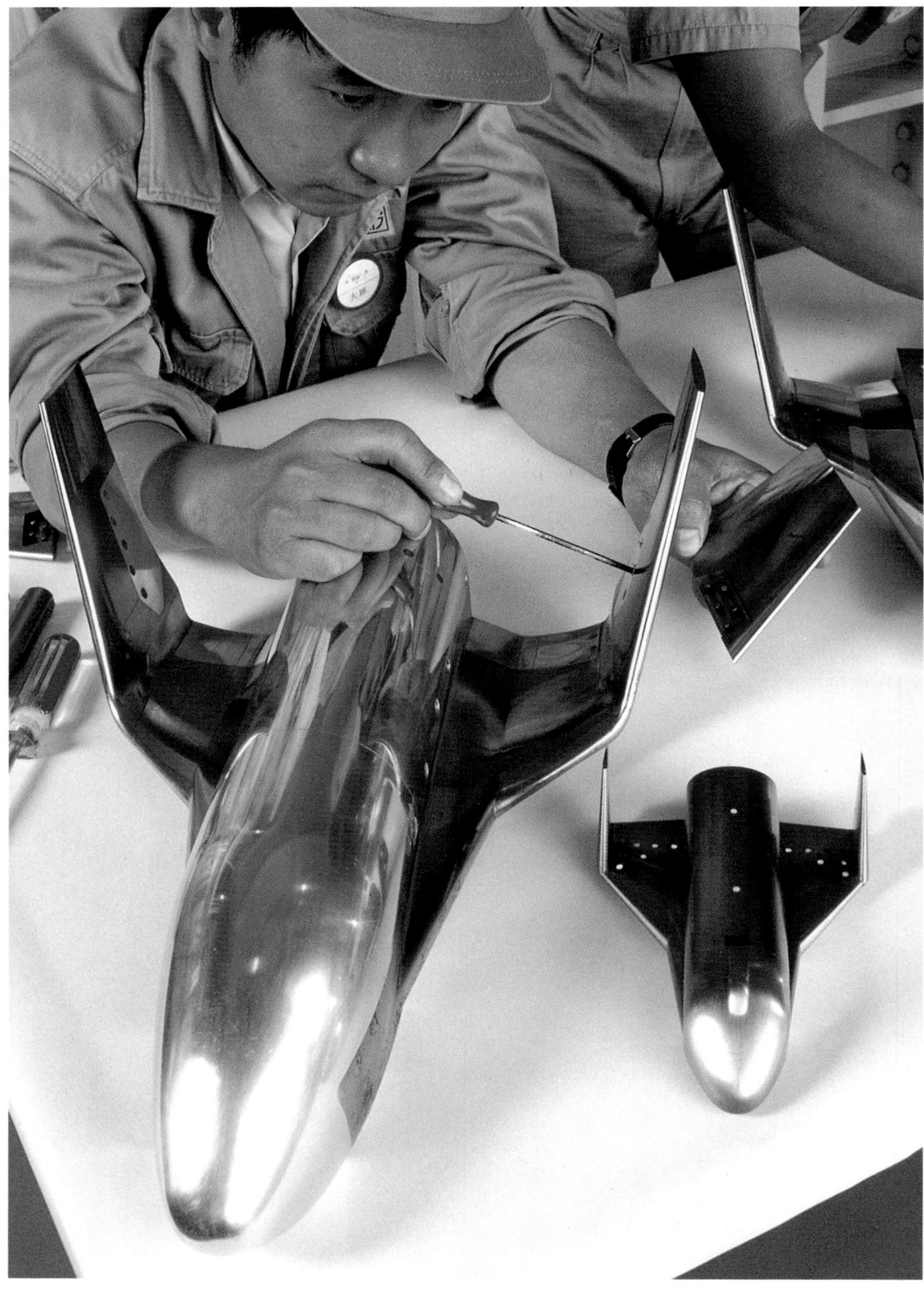

Launching Ahead

An important challenge for the coming century will be the development of low-cost transportation to and from orbit. It is still not clear what form this will take. Some theorize that gravitational forces can be harnessed and reversed; others propose that something like ski-lift cables can connect the earth to satellites in orbit, providing a pathway for an elevator into space.

One day a revolutionary new propulsion system could make rocket technology seem as old-fashioned as the horse and buggy. But until that occurs, the nations of the world will continue the race to improve upon existing rocket-powered systems. The Soviets' new high-lift Energia booster may soon be matched by America's unmanned Shuttle-C, or by the all-new Advanced Launch System (ALS).

The European Community is building a mini-shuttle large enough for three astronauts (top left). Called Hermes, the space plane will provide Europe with independent access to Soviet and American space stations. It will be launched aboard Ariane V using a large new rocket engine, the Vulcain (bottom left).

Japan is moving ahead on a crewless space plane of its own, called Hope (above). These wind tunnel models are used to test the craft's performance at hypersonic speeds. Designed for launch by Japan's new H-2 booster, Hope will be used to recover precious new products that can only be manufactured in orbit.

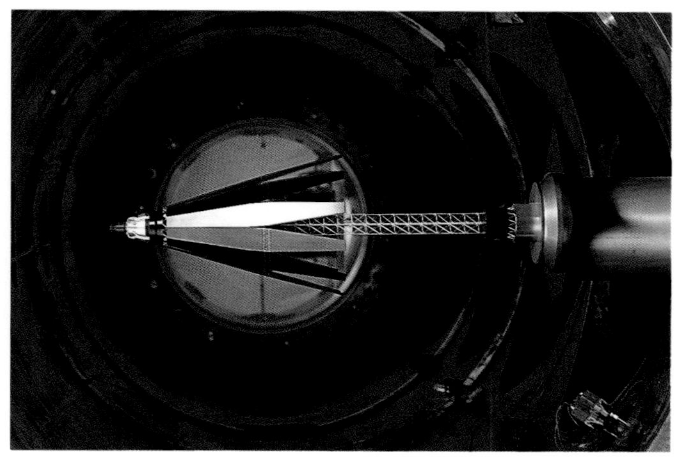

The United States has on the drawing board several improved versions of the current shuttle. Since the Challenger accident, NASA has come to realize that unmanned boosters still provide the safest and most economical way to launch cargo into orbit. Hence a new shuttle would probably carry less payload than current orbiters, but more astronauts. The "Shuttle II" modeled above will use expendable wing tanks and boost eleven people into orbit. With the experience gained from the original orbiters, an improved system promises new levels of safety and economy.

NASA is exploring unconventional propulsion as well, including fuel-efficient nuclear-electric ion drives that can deliver continuous low-level thrust for months on end. The electricity will be supplied by nuclear reactors like the SP-100 (top right).

Finally, the old reliable Titan III (bottom right) has been upgraded, and is being recalled by NASA as a booster for interplanetary probes like the Mars Observer. For the first time, America's Titan, Atlas, and Delta rockets are all available for hire by other nations.

NASA Ames Research Center
Moffett Field, California, U.S.A.

Early in the 21st century a new way of reaching orbit could leapfrog traditional rocket technologies. The U.S. version is called the National Aero-Space Plane, or NASP—a hybrid between a supersonic plane and a spacecraft. NASP will use a combination of jets, ramjets, hypersonic scramjets, and traditional rockets to loft the entire vehicle into orbit. It is designed to take off and land using conventional runways at any major airport. A prototype, called the X-30, is scheduled to fly in the late 1990s. Similar projects are under way in Europe, Japan, and the Soviet Union.

NASP technology demands new ways of designing and testing hypersonic airframes and engines, since conventional wind tunnels can't simulate near-orbital speeds in the upper atmosphere. Supercomputer simulations are used instead. At right, NASA scientist Scott Taylor uses a Cray-II computer to plot pressures and heat on the surface of one possible NASP design as it travels at 25 times the speed of sound.

Another facility at NASA Ames, the Advanced Concepts Flight Simulator, is used to design and test cockpits and instrumentation for the NASP. Control panels will feature fewer dials and more computer screens, like the one "flown" at far right by Bob Shiner.

SURFACE POSITION
TRANSPONDER
COM NAV FREQUENCY
SELCAL

NASA John F. Kennedy Space Center Florida, U.S.A. •

Each day in space, an astronaut consumes about three pounds of food, four pounds of water, and several more pounds of oxygen. Packing enough supplies for a two- or three-year mission to Mars is impractical. Experiments conducted in both the Soviet Union and the United States suggest an alternative—the "closed-loop life-support system"—which mimics the chemical balance between plants and animals on earth.

In an airtight chamber (above), NASA biologists are developing procedures to recycle air, food, and water. Just by breathing, astronauts will supply enough carbon dioxide to sustain a collection of plants—which in turn will give off enough oxygen to sustain the astronauts. Human digestive waste will nourish plants and algae that later become food.

When the project succeeds, a fully operational flight module will be added to space station Freedom. In addition to taking some of the burden off the mechanical life-support system, these techniques may offer spiritual and psychological benefits—the presence of living plants is expected to make life in space much more agreeable.

Biosphere II Test Module Oracle, Arizona, U.S.A. •

When eight human test subjects enter Biosphere II through an air lock and bolt the hatch behind them, they will isolate themselves from the outside world as certainly as if they were on Mars. For two years thereafter, the Biospherian experimenters will be sealed within an airtight, 3.15-acre glass-walled microcosm of the original Biosphere, better known as Mother Earth.

The project promises to develop techniques for self-supporting spacecraft, and for bases on the moon and on Mars. Since everything within Biosphere II's walls—including the air—will be recycled, and no resources except sunlight will be introduced from outside, the project also promises valuable insight into means of protecting and nurturing earth's fragile ecological system. This advanced test of the closed-loop life-support concept is a private enterprise of Space Biosphere Ventures.

One of the biggest challenges is to determine which blend of plants and animals will sustain a closed system. Thousands of species will inhabit various ecological zones including marshland, savanna, desert, a tropical rain forest, an ocean eight meters deep, and a four-story home for the crew. The concept has been tested in the prototype Biosphere module at right and in laboratory exercises.

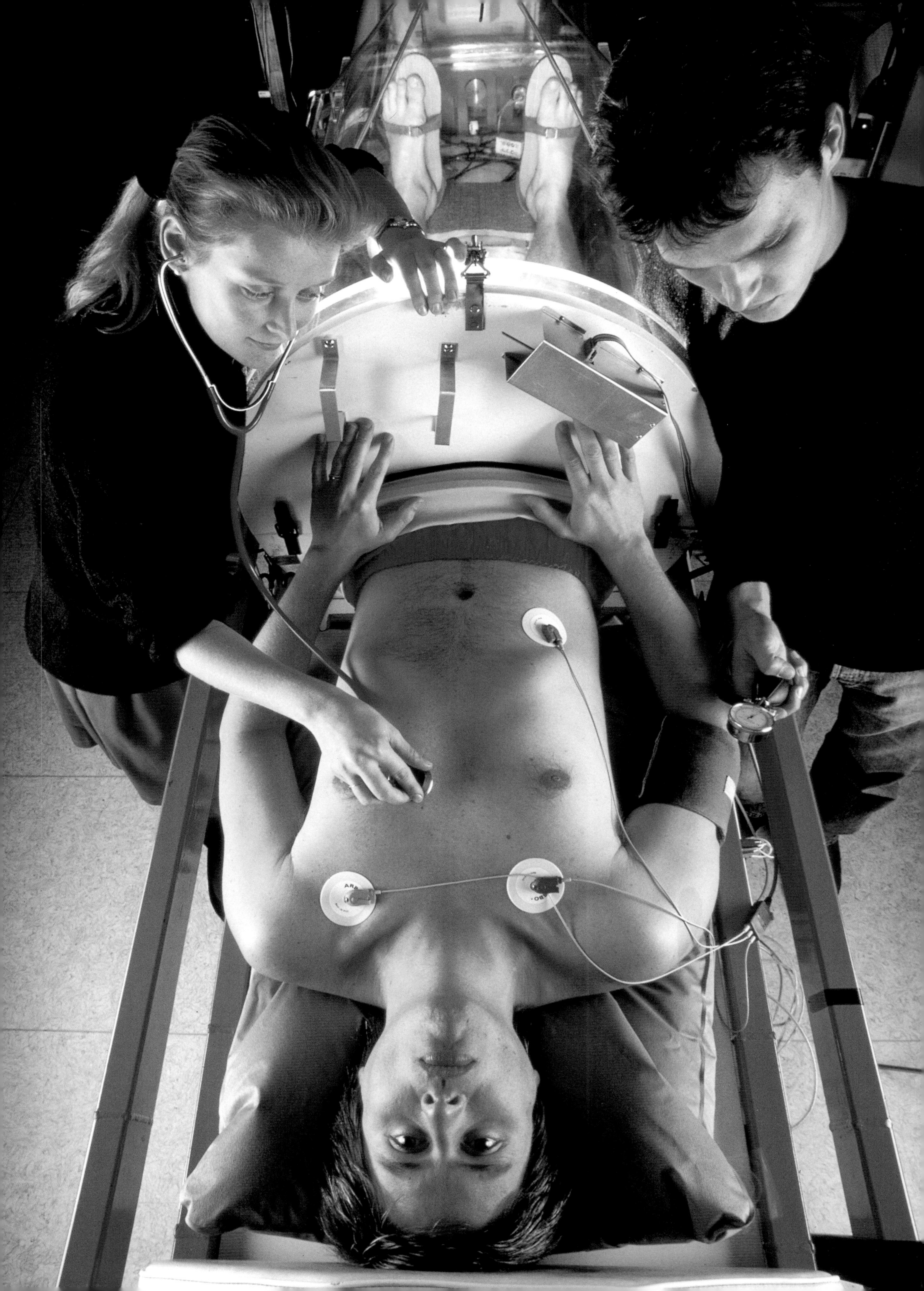

USS Alabama
Pacific Ocean

(Previous pages) Long periods of isolation and confinement in a spacecraft can cause psychological problems. For example, the crew of Skylab 4 is said to have grown testy with ground controllers during its 84-day mission, and Soviet crews lost the motivation to work after months aboard Mir.

To understand what makes a crew happy, harmonious, and productive during lengthy tours of duty, NASA has studied living conditions both at isolated bases in Antarctica and aboard nuclear submarines. Shown here is the berthing area of a Trident sub, which stays at sea for 70 days at a time. The configuration of bunks and personal space is similar to the arrangement planned for the space station Freedom.

DLR Headquarters
Cologne, Federal Republic of Germany

Extended periods of weightlessness, whether in orbit or on a trip to Mars, can lead to serious physiological problems as well. Blood volume and red-cell content shift to undesirable levels, muscles atrophy, the heart shrinks, and bones lose calcium and become brittle.

At left, German astronaut Ulrich Walter undergoes a simulation of returning to earth after a lengthy stay in space. The lower half of his body is inside an airtight glass chamber in which the air pressure has been lowered. Blood rushes to his legs, causing faintness as his brain struggles to get the oxygen it needs. Astronaut candidates who don't demonstrate resistance to the dizzying effects of this experiment are necessarily disqualified from further training.

Above, German scientists prepare for an upcoming Spacelab mission on board NASA's space shuttle. Dr. Heike Walpot exercises on a stationary bicycle, which has been shown to lessen the effects of zero-g by keeping the heart muscles strong. If humans can't learn how to adapt to weightlessness, Mars missions may have to create artificial gravity by twirling two sections of a spacecraft around each other.

NASA Johnson Space Center
Houston, Texas, U.S.A.

Starting in the mid-1990s, many of the questions that need to be answered before we travel back to the moon and on to Mars will be addressed by scientists and engineers aboard space station Freedom. Adaptation to weightlessness and isolation, the manufacture of products in the microgravity environment, the structural integrity of the station during lengthy times in space, and the spirit of international cooperation will all be tested in orbit. Habitation modules will be linked with labs owned by the U.S., Japan, and Western Europe through interconnecting "nodes" like this one. With acrobatic leaps, space station decorators Nick Pausback and Laurie Weaver demonstrate what life will look like in orbit, where there is no "up" or "down."

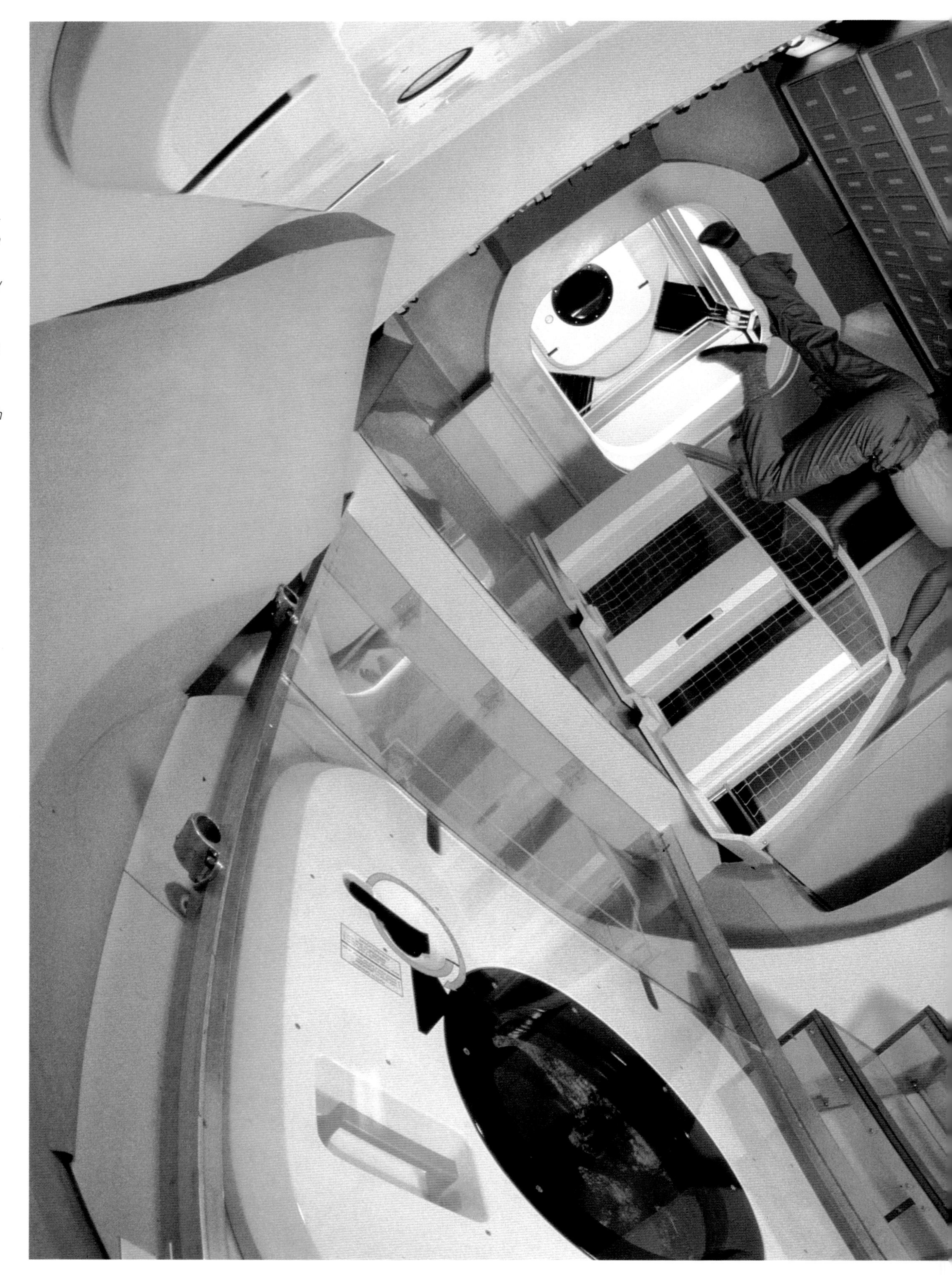

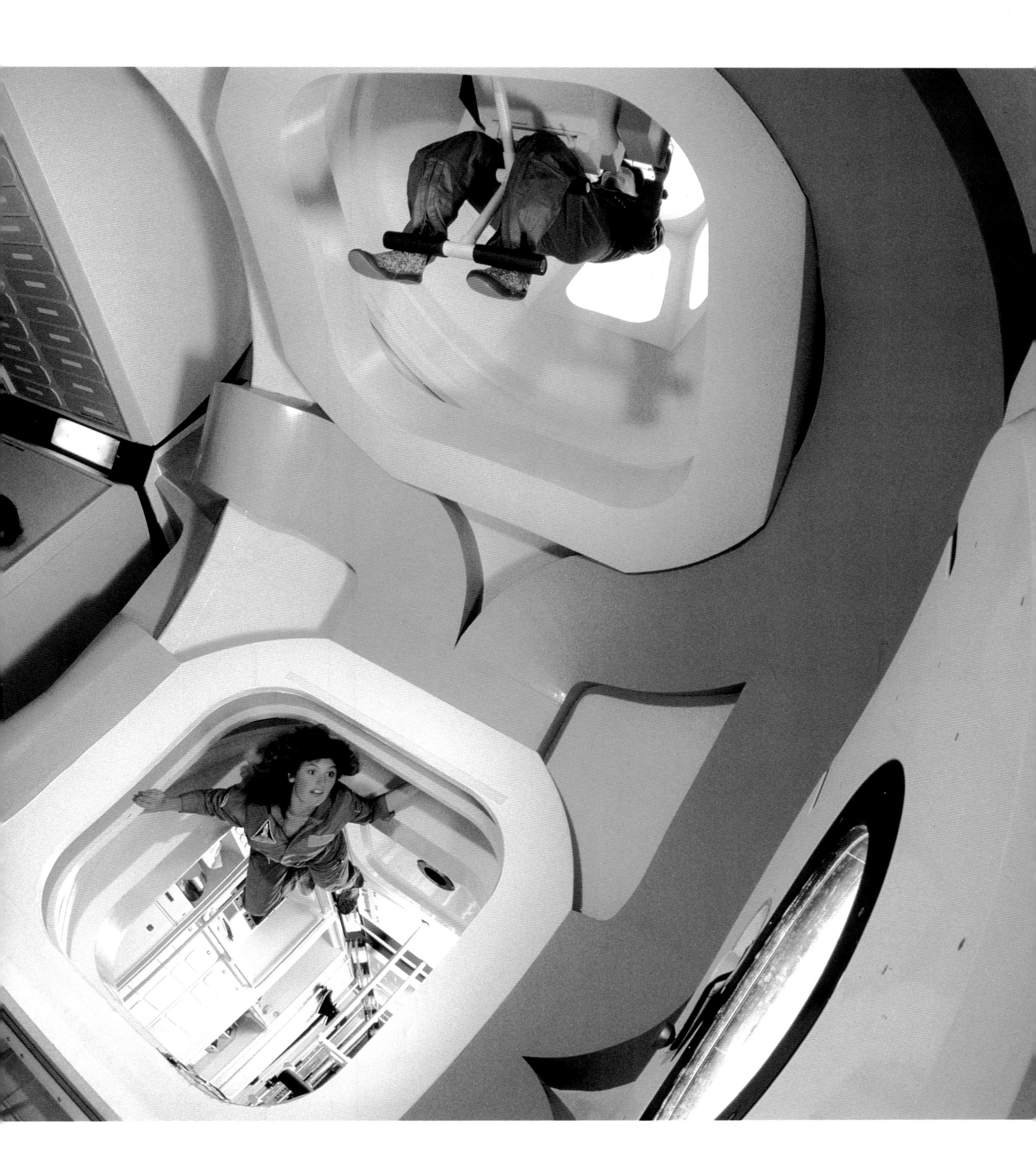

NASA Johnson Space Center
Houston, Texas, U.S.A. ●

(Near right) New space suits are being designed for the 21st century. Here, engineers assess a prototype called the ZPS Mark III, which uses higher internal air pressures than in the past, eliminating hours of "pre-breathing" previously required to avoid getting the bends.

NASA KC-135 Weightlessness Research Plane
Over the Gulf of Mexico

For the space station, NASA plans to introduce a full-body weightless space shower (far right), seen here during a trial run aboard a jet plane known as the "Vomit Comet". The plane follows a roller-coaster flight plan to produce 30-second periods of zero-g. Psychologists say that keeping clean in space—a difficult task in the past—will make for happy crews.

NASA Ames Research Center
Moffett Field, California, U.S.A. ●

(Following pages) This novel prototype spacesuit, the AX-5, uses interchangeable sizing rings between the joints so that one size fits all, an important consideration on space station missions with rotating crews. The AX-5's aluminum shell is expected to provide better protection against micrometeoroids than the rubber-and-fabric suits of the past. The suit is modeled by creator Vic Vykukal, dubbed the "Bill Blass of space-suit design" by Life Magazine.

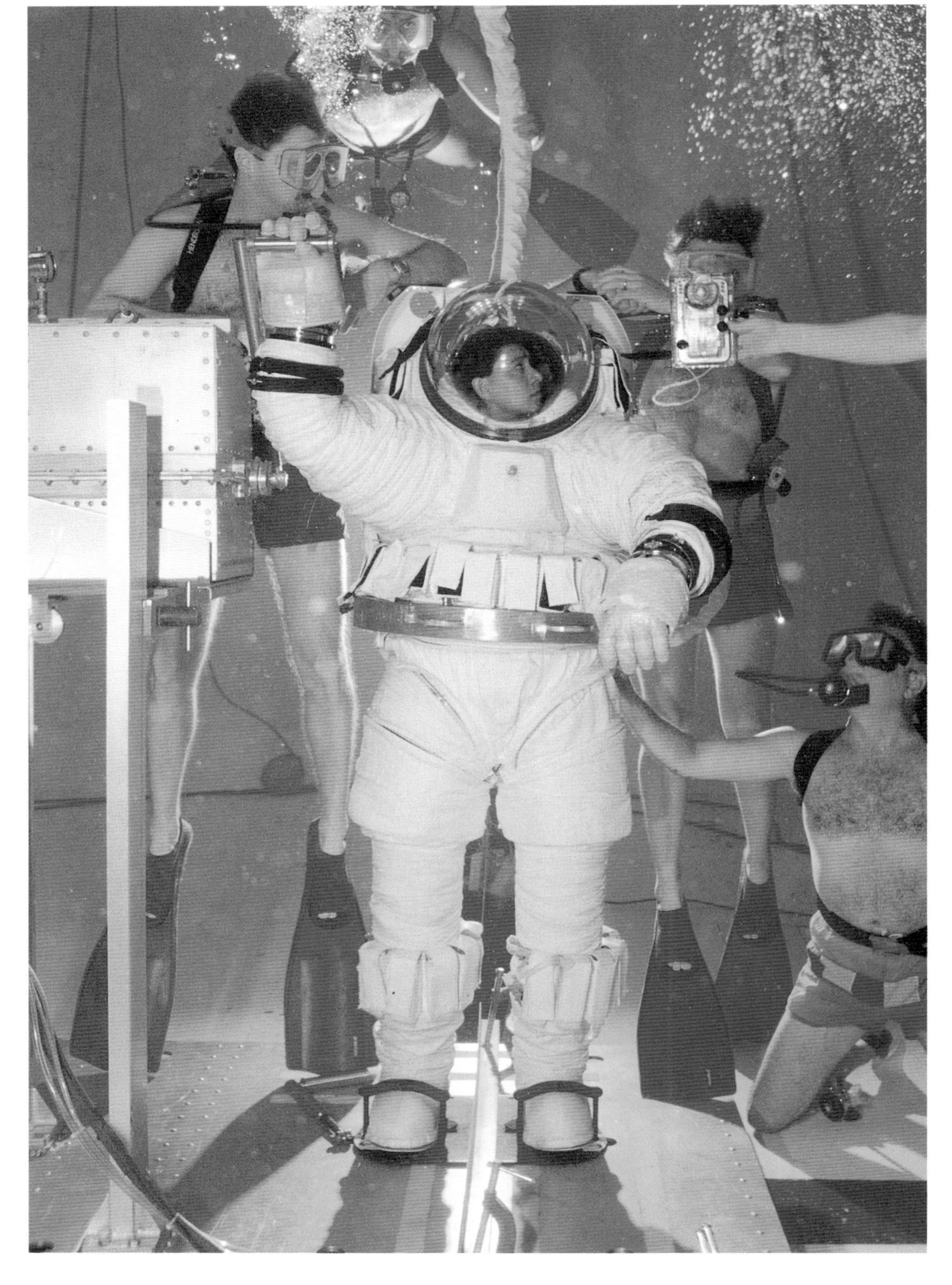

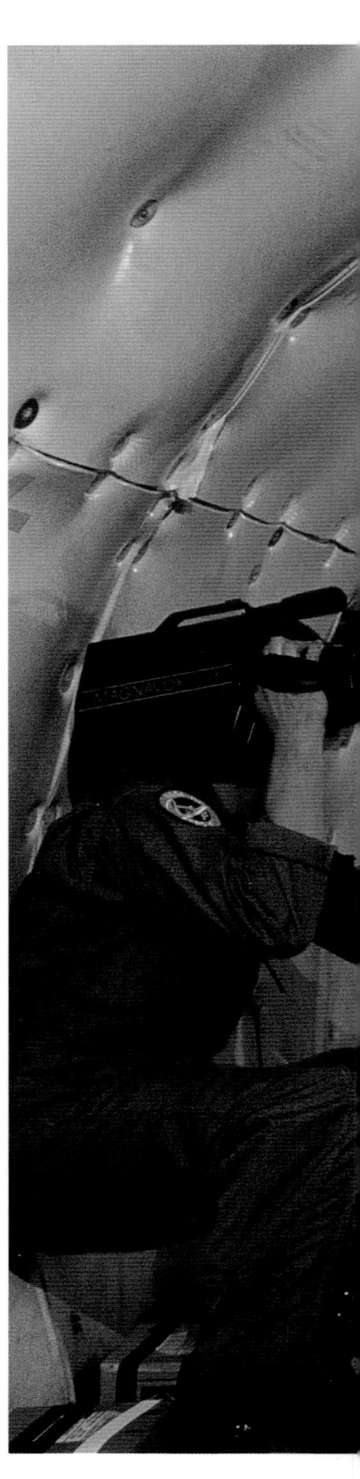

ACKNOWLEDGMENTS

More than any other person, Raymond H. DeMoulin of the Eastman Kodak Company made the full scope of this book possible. Kodak's generous support, and Ray's faith in the project, enabled me to travel the world for a year and chase my dream to completion.

Others in Kodak's Professional Photography Division who helped include Ann Moscicki, Marianne Samenko, Doug Putnam, and those who answered my technical questions over the years: John Altberg, Don Stang, Jeff Peterson, and Robert Shanebrook.

Mike Gentry at NASA's Johnson Space Center and Jurrie van der Woude at the Jet Propulsion Laboratory made significant contributions and patiently indulged me while I gathered and researched this book's NASA pictures.

Several magazine and book editors furthered my career "way back when," and have been important friends during good times and bad: Mary Dunn, Tom Kennedy, Michele Stephenson, John Loengard, Scott DeGarmo, Caroline Despard, Toni Burbank, Rick Smolan, and David Cohen. Some of these people offered photo assignments that later became the seeds for this book.

The Nikon Corporation has supplied invaluable assistance over the years. I'd like to thank Mike Phillips, Kimio Shioiri, Richard LoPinto, Bill Pekala, Sam Garcia, and especially Scott Andrews.

Photo assistants in all corners of the world, too numerous to list here, have shared these experiences with me, sometimes under difficult conditions and often in the middle of the night. My parents and sisters were the first, and I am grateful to all.

Many, many others have helped this book evolve during the past 20 years, and I ask forgiveness if any names have been omitted in the process of compiling the following lists. You know who you are, and I thank you.

My editor Bruce Gray, the book's talented designer Jennifer Barry, editors Bill Messing and Michael Lemonick, and many others among the Collins staff have been helpful and supportive—a joy to work with as the project reached its final stages. Robert Cavallo heard the confessions of all. And Jain Lemos of Starlight Photo Agency has worked especially hard during the past year, facilitating the completion of this remarkably complex project—made all the more exciting by the timing of the California earthquake of 1989, and by our office's location in the hard-hit Marina District of San Francisco.

—Roger Ressmeyer, April 1, 1990

ADDITIONAL PHOTO CAPTIONS

Front Cover: Las Campanas Observatory, Chile
With this telescope, Ian Shelton discovered Supernova 1987A. The supernova and its home galaxy, the Large Magellanic Cloud, are visible just off the front end of the telescope. The camera for this photograph was provided by the Steward Observatory Mirror Lab.

Pages 6-7: Mauna Kea Summit, Hawaii, U.S.A.
The stars trace circles around the North Celestial Pole as the earth spins during a two-hour exposure of the Canada-France-Hawaii Telescope.

Pages 8-9: Palomar Observatory, California, U.S.A.
Palomar Mountain's giant five-meter (200-inch) telescope was the world's largest from 1949 to 1976. Its protective dome appears transparent because it rotated during this exposure.

Pages 10-11: Ames-Dryden Flight Research Facility, California, U.S.A.
Space Shuttle Enterprise, a prototype that never orbited, during a presidential celebration after the fleet's final test flight.

Pages 12-13: Low Earth Orbit, Apollo-Soyez
American astronaut Deke Slayton and Soviet cosmonaut Aleksey Leonov embrace after docking their spacecraft during an historic mission in 1975.

Pages 14-15: Kennedy Space Center, Florida, U.S.A.
The first shuttle flight after a 32-month hiatus following the Challenger disaster (in which seven astronauts perished just moments after lift-off) captivated the world.

Pages 16-17: Tanegashima, Japan
A Japanese H-1 rocket bearing an advanced weather satellite thrusts toward orbit with the help of six "strap-on" solid fuel engines.

Pages 18-19: Sea of Serenity, The Moon
Geologist Harrison Schmitt, the first working scientist to visit the moon, studies a giant boulder in the Taurus-Littrow area.

Pages 22-23: The Cone Nebula in Monoceros, Milky Way Galaxy
A gas and dust cloud is brilliantly illuminated by young stars. In the foreground, a darker, cone-shaped cloud obscures the scene behind it.

Pages 32-33: Las Campanas Observatory, Chile

Pages 64-65: The Center of the Milky Way Galaxy.
Photo by Roger Ressmeyer with camera by Steward Observatory Mirror Lab.

Pages 112-113: Kennedy Space Center, Florida, U.S.A.
Lightning strikes near the Space Shuttle on launch pad 39A.

Pages 160-161: Jupiter Photographed by Voyager 1
Galilean moons Io and Europa orbit past Jupiter's Great Red Spot.

Pages 174-175: NASA Johnson Space Center, Texas, U.S.A.
Simulation of a final approach to Mars.

Pages 200-201: Las Campanas Observatory, Chile
Ian Shelton takes his last look at Supernova 1987A through a pair of binoculars before returning to the University of Toronto in July, 1987.

Pages 202-203: Roque de los Muchachos Observatory, La Palma, Canary Islands, Spain
The constellation of Orion the Hunter rises over the dome of the William Herschel Telescope. In Orion's "sword," hanging to the right of his belt, can be seen the Orion Nebula.

Pages 204-205: The Orion Nebula
A close-up reveals intricate detail in this glowing cloud of interstellar gas, a birthplace for young suns and solar systems.

Page 208: Kennedy Space Center, Florida, U.S.A.
Just seconds after liftoff in March of 1989, space shuttle Discovery rolls and pitches toward its proper orbital trajectory.

SPACE PLACES

Concept, Photography, Photo Gathering, Text
Roger Ressmeyer

Project Manager/ Assignment Editor
Jain Lemos

Coordinator, Starlight Photo Agency
Sandra Carver

COLLINS PUBLISHERS

President
Bruce W. Gray

Design Director
Jennifer Barry

Managing Editor
Bill Messing

Production Director/ Design Assistant
Stephanie Sherman

Science/Technical Editor
Michael Lemonick

Production Assistant
Diana Jean Parks

Publicity Director
Patti Richards

Sponsorship Director
Cathy Quealy

Sales Director
Carole Bidnick

General Manager
Jennifer Erwitt

Sponsorship Manager
Blake Hallanan

Business Manager
Peter Smith

Office Manager
Linda Lamb

Staff Editor
Kate Kelly

Senior Accountant
Jenny Collins

Production Coordinator
Kathryn Yuschenkoff

Editorial Assistants
Brian Hajducek
John Clay Stites

Staff Designer
Kari Ontko

Sponsorship Assistant
Monica Baltz

Sales Assistant
Maria Hjelm

Administrative Assistants
James Kordis
Scott MacConnell
Jill Stauffer

Researcher
Virginia Rich

Proofreaders
Charles Hopkins
Jonathan Schwartz

Attorneys
Coblentz, Cahen, McCabe & Breyer, San Francisco

SPONSORS

Eastman Kodak Company
Nikon, Inc.
International Business Machines Corporation

MAJOR CONTRIBUTORS

Andrews Photographics
Anglo-Australian Observatory
The Astronomical Society of the Pacific
Best Photo Labs
The B.O.S.S. Software
California Academy of Sciences
Life Magazine
Meade Instruments Corporation
Media Services Corporation
National Aeronautics & Space Administration
National Geographic Magazine
National Space Development Agency of Japan
Nikon Corporation, Japan
Pallas Photo Labs, Inc.
Smithsonian Magazine
Steward Observatory Mirror Lab
Time Magazine

CONTRIBUTORS

Academy of Sciences of the U.S.S.R.
Aerospatiale
Air & Space Magazine
Amateur Astronomers Association of New York City
Ames Pressure-Suit Staff
Arianespace, Inc.
Ball Aerospace Systems Group
Boehringer Mannheim Corporation
British Airways
Carnegie Institution
Centre National d'Etudes Spatiales
Comet Corporation
The Darkroom
Discover Magazine
Dyna-lite
English Heritage
European Space Agency
German Aerospace Research Establishment
Glavkosmos
Global Ocean Chlorophyll Project
Institute for Astronomy
Intourist
Lockheed Missiles and Space Company, Inc.
Long Island Lutheran High School
Lumicon
Martin Marietta Corporation
Max Planck Institute
MGI Studio
The Ministry of Aero-Space Industry, People's Republic of China
The New Lab
Newsweek Magazine
New York Times Magazine
Novosti Press Agency
Rockwell International
Royal Observatory Edinburgh
San Francisco Examiner Image Magazine
Science Digest Magazine
Sky & Telescope Magazine
Space Commerce Corporation
U.S. Geological Survey
U.S. Naval Research Laboratory
U.S. Space Camp
Virginia Roth's Scientific Expeditions, Inc.

Dai Nippon Printing Co., Ltd.
Ryo Chigira
Fujio Ojima
Akira Ishiyama
Kikuo Mori
Mitsuo Gunji
Yoshio Akasaka
Yoshinori Katoh

AUTHOR'S SPECIAL THANKS

Buzz Aldrin
Lois Aldrin
Roger Angel
Lonny Baker
Kelly Beatty
Anatoly Bogomolov
Thomas Canby
Robert Cavallo
Bruce Chinn
Rich Clarkson
Jean Clough
François Colas
Michael Collins
Wendy Determan
John Diebel
Bill Douthitt
Arnold Drapkin
Alan Dressler
Vladimir Dzhanibekov
Mary Dawn Earley
Andy Fraknoi
Wilbur Garrett
Curt Grosjean
John Gustafson
Don Hall
Steve Hammond
Stan Hays
Judy Hilsinger
Peter Howe
Robert Kirshner
Michael Lemonick
Debra Lex
David Malin
Jack Marling
Sandi Mendelson
Gary Morrison
Don Moser
Karen Mullarkey
Stephen Nesbitt
James Oberg
Keiko Ohmura
Richard Pedrelli
Charles Redmond
Ed Rich
Felix Rosenthal
Mike Ross
Kathy Ryan
Jon Schneeberger
Diane Stanley
Richard Truly
Vic Vykukal
Eugene Zykov

SPACE PLACES PICTURE AGENTS

North America
Jain Lemos, Director
Starlight Photo Agency
(a division of Roger Ressmeyer Photography, Inc.)
2269 Chestnut Street #400
San Francisco, California 94123
Telephone: (415) 921-1675
Fax: (415) 921-2006
Telex: 650-259-4088-MCI

France
Annie Boulat
Cosmos
56 Boulevard de la Tour Maubourg
75007 Paris
Telephone: (1) 4705-4429
Fax: (1) 4705-4205
Telex: 203085

Holland, Belgium, Luxembourg
Ellen Van De Graaf
ABC Press Service
O. Z. Achterburwal 141
1012 DG Amsterdam
Telephone: (2) 024-9413
Telex: 13498

Italy
Grazia Neri
Via Parini, 9
20121 Milan
Telephone: (2) 650-832
Fax: (2) 659-7839
Telex: 312575

Japan
Bob Kirschenbaum
Pacific Press Service
CPO 2051 Tokyo
Telephone: (3) 264-3821
Fax: (3) 264-3899
Telex: 26206

Scandinavia
Anette Schneider
IFOT
Rosenvaengets Alle 37
DK 2100 Copenhagen
Telephone: (3) 138-6111
Fax: (3) 543-1611
Telex: 22954

Spain
Alfonso Gutiérrez Escera
A.G.E. FotoStock
Buenaventura Muñoz, 16, Entlo.
08018 Barcelona
Telephone: (93) 300-2552
Fax: (93) 309-3977
Telex: 51743

United Kingdom
Michael Marten
Science Photo Library
112 Westbourne Grove
London W2 5RU
Telephone: (1) 727-4712
Fax: (1) 727-6041

West Germany
Margot Klingsporn
Focus
Moorweidenstr. 34
2000 Hamburg 13
Telephone: (40) 44-3769
Fax: (40) 45-6804
Telex: 216-4242

FRIENDS, ADVISERS & CONSULTANTS

Victor Abalakin
Pam Abramson
Tony Acevedo
Joe Acree
Bruce Adams
Eddie Adams
Bob Adelman
Thelma Altemus
Walter Alvarez
Dorothy Affa Ames
Laura Amor
Peter Anderson
Isabelle Anglin
Wayne Annala
David Arnett
Yoshihiko Asai
Ron Bailey
Bobbi Baker-Burrows
Marty Balin
Devi Baptiste
Brigitte Barkley
Carole Baron
David Barry
Ken Barton
Cathy Baskin
Lee Battaglia
Henry Battjes
Alan Bean
Garnet Beard
Lance Beauchamp
Bob Beckmann
Behnam
John Bergez
Linda Billings
Lawrence Blakee
Richard Boeth
Rene Bosson
Beatrice Bowles
Cynthia Bowman
John Boyd
Richard Bradford
Frisch Brandt
Bill Brence
Kathryn Bretscher
Manuel Bretscher
Lois Briggs
Dan Brocious
David Brown
Jane Brown
John Brown
Robert Brucato
Paul Buchanan
Sheri Buckaloo
Bruce Buckingham
Philippe Buffet
Sandra Burton
Paul Butterworth
Michael Byrnes
Sean Callahan
Tom Campbell
Duncan Campbell-Wilson
Russell Cannon
Frank Card
Karen Cardell
Ted Carmack
Jeff Carr
Steve Carroll
Jane Carruth
Lloyd Carter
Eric Cerf-Mayer
Eugene Cernan
Fred Chaffee
Craig Chaquico
Jean-Louis Charles
Elizabeth Cheng-Krist
Yvonne Clearwater
Aaron Cohen
Sue Cometa
Shirley Compard
Nadine Condon
Susan Contois
James Cook
Guy Cooper
Elizabeth Cope
Carolyn Cox
Margo Crabtree
Ctein
Cui Shizhu
Gillian Cutner
Bill Daly
Jeanne David
Michael Davis
Billie Deason
Judy DeHaas
Mario de Lepine
Bill Del Giudice
George Diller
John Dobson
Liz Doherty
Michael Downey
David Drachlis
Thomas Dreschel
Art Dula
Sheri Dunnette
Kathleen Dyhr
Gerald Ebker
John Echave
Sue Eley
Ed Erickson
Michael Evans
Robert Evans
Susan Evans
Nancy Faber
Don Fail
Dianne Feinstein
Gene Feldman
Carolyn Fields
Scott Fisher
Elizabeth Forst
Lisa Fowler
Nancy Frye
Albert Galeev
Rafael Garcia
Lucy Garrett
Mark Gartland
Kelly Gatlin
Brigitte Gauthier
Jan Gauthier
Isabel Geffner
Karen Gerold
Hoot Gibson
Carl Gillespie
Peter Gillingham
Owen Gingerich
Bill Giordano
Penni Gladstone
Deborah Godfrey
Rae Ellen Godfrey
Rick Gore
Jennifer Gotti
Michael Greenisen
Clifford Grobstein
Ruth Grobstein
Susan Griffin
Cyndi Griffith
Paul Grover
Ernie Haberer
Magne Hagstrom
Barbara Hall
George Hall
Del Harding
Woody Harrington
Ed Harrison
John Harvey
Louis Haughney
Stephen Haupt
Steven Hawley
Walter Heard
Tina Helsell
William Hempel
Kyle Herring
Chuck Herron
Matt Herron
Gregory Hess
Fred Hill
Eijiro Hirohama
Pam Hoffenberg
Nancy Hoffman
Sam Hoffman
Danielle Hofstedt
Holly Holden
Robert Holmes
Richard Howard
Richard Howenstine
Peggy Huang
Ramon Huidobro
James Humphries
Vern Iuppa
Susan Ives
Jim Jackson
Ken Jacobs
Donald James
Mark Jansen
David Jauncey
Tom Johnson
Albert Jones
Jeffrey Jones
Ken Jones
William Jones
Nancy Kahan
Devyani Kamdar
Marita Kankowski
Paul Kantner
Horst Uwe Keller
Maureen Kelly
Karen Kenagy
Eamon Kennedy
Ayumi Kimura
Jane Kinne
Douglas Kirkland
Christa Klein
Leonard Klein
Naomi Kleinberg
William Knott
Kent Kobersteen
Hermann Kochan
Charles Kogod
Tony Kolz
Arthur Koski
Joe Kosmo
Volker Kratzenberg-Annies
Stanley Krippner
Bob Kubara
James Kuhn
Irina Kulinich
Judy Ladendorf
Pietra LaRotonda
April Lasher
Ron Laub
Scott Lawrence
Billie Jean Lebda
Chong Lee
Ken Lee
Jack Lemos
Toni Lemos
Jean Lentignac
J.P. Leventhal
Valentin Lipovetsky
Rudy Lopez
Nancy Lovato
Terry Lowenthal
Richard Luce
Luo Tao
Michelle Lyle
Gary MacDonald
Don Machholz
Shirley MacLaine
Robert MacMillin
Robert Madden
Nancy Madlin
Judi Magann
Cheryl Magazine
Susan Magrino
Lisa Malone
Janice Maloney
Dick Manchester
Jude Manna-Sullivan
Helen Marcus
M.C. Marden
Gene Marianetti
Brian Marsden
Blaine Marshall
Leonard Martin
Elizabeth Martinez
Harold Masursky
Jeannie Matthews
Yoni Mayeri
Lyndine McAfee
Barbara McConnell
John McCormick
Susan McCurry
Gary McDonald
Bruce McElfresh
Donna McKinney
Al McKuen
John McLeaish
Stan Menscher
Roger Meredith
Daniel Metzle
Rachael Miles
Maddy Miller
Peter Miller
Stanley Miller
Virginia Miller
Robert Miserentino
Andrew Mishkin
Edgar Mitchell
Melanie Mitchell
Bill Moran
Caroline Murphy
Kim Murray
Mary Beth Murrill
Mutsuhiko Masuda
Takashi Nagano
Matthew Naythons
James Neihouse
Leslie Neihouse
Jerry Nelson
Amy Note
Cooky Oberg
Hiroyuki Oguma
Orbit & Umbra
Dennis Overbye
Rusty Pallas
Costas Papaliolios
Neil Parker
Sean Parks
Jay Pasachoff
Andrew Patnesky
Nick Pausback
Charles Pellerin
Charlene Peng
Mary Ann Peto
Patricia Phillips
Keith Pierce
Gerard Pinneberg
Kimberly Pisciotta
Jim Plumeri
Jessica Portnoy
Byron Preiss
Sandra Preston
Gary Price
Carol Prince
Tom Prince
Garry Qualls
Richard Rath
Ruth Ravenel
Steve Raymar
Charles Reich
Tom Reid
Alon Reininger
Harold Reitsema
Michael Remer
Faith Ressmeyer
Georgia Ressmeyer
Henry Ressmeyer
Ruth Ressmeyer
Diana Richmond
Guenter Riegler
John Riley
Patrick Rizzo
Bill Robbins
Cindy Robinson
Leif Robinson
Robert Roemer
Michael Rogers
Terry Romero
Jerry Ross
Evelyn Roth
Virginia Roth
Christophe Rothmund
Barbara Rowell
Galen Rowell
Al Royce
Mario Runco
Ute Runkel
Anne Russell
Mark Rykoff
Roald Sagdeev
Laura Sammarco
Claude Sanchez
Jacky Sarti
Carolyn Savarese
Henry Scanlon
Hans Schlegel
Harrison Schmitt
David Schonauer
Barry Schroder
Marie Schumann
Hans-Emil Schuster
Paul Schwartz
Rolf Schwartz
Karen Schwartzman
David Scott
Lurton Scott
Pete Sears
Margaret Sedgwick
Michael Sela
Paul Sewell
Ian Shelton
Eleanor Sherman
Robert Sherman
Noboru Shimizu
Steve Shindler
Robert Shiner
Eugene Shoemaker
Si Xuewu
Paul Siemers
Mike Simmons
Robert Sims
Robert Sirota
Victoria Sirota
Grace Slick
Brad Smith
Harlan Smith
Kimberley Smith
Marck Smith
Nancy Smith
Thomas Smith
Leslie Smolan
Patsy Somlo
George Sonneborn
Frank Soper
David Spitzler
Harold Stall
Joe Stancampiano
Maria Stenzel
Sue Stephensen
Olivia Stewart
Ed Stone
Gary Stone
John Stovell
Ted Streshinsky
Peter Strittmatter
Vic Studer
Barry Sundermeier
Jody Swann
Paul Swanson
Kosuke Tago
Jon Tandler
Tom Tarnowsky
Bill Taylor
Zaki Tewfik
Robert Thicksten
Gerhard Thiele
Pierre Thuot
Saburo Tojyo
Barbara Townsend
Danielle Traina
David Travis
Frederick Trinklein
Jeanne Trombly
George Ulich
Susan Vacurevich
Charlene Valeri
Martha Vanderkolk
Lisa Vazquez-Morrison
Manny Virata
George von Kantor
Dianna Waggoner
J.D. Wahiche
Roy Wallis
Heike Walpot
Eileen Walsh
Ulrich Walter
Dan Walters
Sharon Walters
Wang Zhen
Douglas Ward
Jack Warren
Laurie Weaver
Bruce Webbon
Richard Weisgrau
Brian Welch
Dennie Welsh
Margaret Wenger
Paul Wenger
James Wetherbee
Maria Wilhelm
Bob Williams
Kathe Willis
Ann Winblad
Sandie Witbeck
Nick Woolf
Stan Woosley
Steve Wozniak
Jan Wrather
Kenji Yagihara
Kenji Yamaguchi
Victor Yerin
Dick Young
Ray Yost
Cathy Zier
Mary Zisk